Sea Ice Image Processing
with MATLAB®

Signal and Image Processing of Earth Observations Series

Series Editor

C.H. Chen

Published Titles

Sea Ice Image Processing
with MATLAB®

Qin Zhang and Roger Skjetne

CRC Press
Taylor & Francis Group
Boca Raton   London   New York

CRC Press is an imprint of the
Taylor & Francis Group, an **informa** business

CRC Press
Taylor & Francis Group
6000 Broken Sound Parkway NW, Suite 300
Boca Raton, FL 33487-2742

First issued in paperback 2020

© 2018 by Taylor & Francis Group, LLC
CRC Press is an imprint of Taylor & Francis Group, an Informa business

No claim to original U.S. Government works

ISBN 13: 978-0-367-57218-1 (pbk)
ISBN 13: 978-1-138-03266-8 (hbk)

Visit the Taylor & Francis Web site at
http://www.taylorandfrancis.com

and the CRC Press Web site at
http://www.crcpress.com

Dedication

This is a really elaborate book for the image processing to extract ice information such as ice concentration, ice floes size distribution, and ice type. To my knowledge, this is the first work that attempts to find the method to obtain ice floe size distribution in a systematic and sophisticated way. Nowadays the ice concentration data on a global scale has become available on a daily basis due to the development of microwave satellite sensors. According to this innovation, it has become possible to monitor the variability of sea ice extent on a global basis. However, it is still a big issue to predict the sea ice behavior in the numerical sea ice model due to the lack of our knowledge about the sub-grid scale information of sea ice. To solve this, the attention to ice floe size distribution has recently been increasing. Not limited to the climate problem, sea ice information on a sub-grid scale is also an important issue to the maintenance of the infra-structure built in the polar ocean. Despite its importance, one of the biggest problems has been the difficulty in image processing analysis. Since sea ice is a complicated matrix, it is not easy to analyze the sea ice images quantitatively. This has hampered our understanding of the sea ice properties on a sub-grid scale. This book aims at solving this problem and has achieved it to some extent by elaborately examining the previous methods using a tank model and improving the so called the gradient vector flow snake algorithm for the real ice conditions. This work certainly contributes to this issue and sheds light on collecting observational data on an operational basis. Therefore, I highly evaluate this work and hope the algorithm developed by the authors will be used to improve our understanding of sea ice properties.

— Takenobu Toyota
Institute of Low Temperature Science
Hokkaido University

Contents

Foreword

The Centre for Research-Based Innovation SAMCoT (Sustainable Arctic Marine and Costal Technology) was established in 2011 by the Research Council of Norway to produce the science needed to safely and sustainably explore and exploit Arctic resources. The center's researchers and PhD candidates are namely investigating topics including how to strengthen ships or stationary structures to protect them in collisions with ice and developing computer models that can predict sea ice and iceberg drift and also help understanding the effect sea ice has on coastal erosion.

The present book on "Sea Ice Image Processing with MATLAB®" by Qin Zhang and Roger Skjetne was written as part of their research at SAMCoT at a time when the Arctic climate change is happening twice as fast as elsewhere leading to retreatment and thinning of the ice cover both in summer and winter. This fact enables more extensive exploitation of the Arctic with renewed interest for offshore developments, deep-sea mining, fisheries, shipping, and ship tourist traffic.

The real Arctic sea ice is by no means a continuous level ice field as assumed by many engineering design codes. Instead it is mainly comprised of discontinuous features of fragmented ice, whether it is naturally broken, e.g. in the marginal ice zone by gravity waves, or artificially broken by physical ice management. Depending on the confinement, ice concentration, and floe size distribution, the governing mechanisms during the interaction between sea ice and marine structures or ships can differ considerably.

The book deals with the topic of image processing for the extraction of key sea ice characteristics from digital photography, which is of great relevance for Arctic remote sensing and marine operations. The developed algorithms enable segmenting ice regions from water regions, detecting ice floe boundaries and floe positions, classifying different forms of sea ice, and estimating ice concentrations and floe size distributions from images of broken ice fields. This is highly relevant for design of structures intended for ice-covered seas in the Arctic as well as for studies in climate change. Full-scale experience, ice tank tests, and numerical models are frequently used when designing marine structures for icy waters. The book by Zhang and Skjetne contributes strongly to provide tools for quantifying the ice environment that need to be identified and also reproduced for such testing. This includes fit-for-purpose studies of existing vessels, new-build conceptual design and detailed engineering design studies for new developments, and studies of demanding marine operations involving multiple vessels and operational scenarios in sea ice.

The research efforts presented in this book can help ensure that these developments do not take their toll on the environment, by giving the industry the knowledge they need to protect vital infra-structure and avoid accidents, as well as being valuable for planning and performing safe marine operations in the Arctic.

Director of SAMCoT   Professor Sveinung Løset, NTNU

Preface

Recently the interest in small scale dynamics and properties of sea ice has increased as a result of more detailed science on Arctic processes and a growing industrial activity in the North. Marine research expeditions to Arctic ice-infested waters typically include many camera-based technologies to document the conditions of the sea ice and environment during the operations. Camera technology is cheap and easily available, and the pictures of the sea ice provide invaluable information on the type and state of the sea ice and the corresponding physical phenomena taking place. In the aftermath, the time and geo-referenced images provide documented evidence of the state of the sea ice at the corresponding time and place.

Image processing of the sea ice images is vital in order to extract, quantify, and numerically process the relevant sea ice parameters and understand the behavior of the sea ice field, especially on a relatively small scale. However, since sea ice is a complicated multi-domain process, it is not easy to analyze the sea ice images quantitatively. The lack of the necessary tools and systematic methods for image processing has hampered the understanding of the dynamical properties of sea ice on a sub-grid scale. Hence, the development of automated computer algorithms for efficient image analysis and accurate extraction of some interesting sea ice parameters from the images is a main objective of this book.

This book focuses on the computer algorithms to process individual sea ice images and video streams of images to extract parameters such as ice concentration, ice floe size distribution, and ice types, which are very important in studies of climatic changes in the polar regions, as well as in various fields of ice engineering.

The Otsu thresholding and k-means clustering methods are employed to identify the ice from the water and thereby to calculate ice concentration. Both methods are effective for the sea ice images in which all the ice is significantly brighter than the water. However, due to the ability of also detecting "dark ice" with brightness close to water, it is found that the k-means method is more effective than the Otsu method for the sea ice images in which a large amount of brash ice and slush exist.

The derivative edge detection and morphology edge detection methods are presented and employed to find the boundaries of the ice floes. These methods are efficient in detecting distinct boundaries in the image, but they have a weakness in detecting closed boundaries of the individual ice floes and to separate the ice floes that are tightly connected. To overcome this problem for the identification of individual ice floes and quantifying ice floe statistics, the watershed transform and the gradient vector flow (GVF) snake algorithm are presented and applied.

In the watershed-based method, the grayscale sea ice image is first converted into a binary image and the distance transform-based watershed segmentation algorithm is carried out to segment the image. A chain code is then used to check the concavities of the floe boundaries. The segmented neighboring regions that have no concave corners between them are merged, so that over-segmentation lines are automatically removed. This watershed-based method is applicable to separate the seemingly connected floes for which the junctions are invisible or lost in the image. Hence, the method can be used for individual ice floe detection, though some over- and under-segmentation will typically be present.

In the GVF snake-based method, seeds for each ice floe are first obtained by calculating the distance transform of the binarized image. Based on these seeds, the snake contours with proper locations and radii are initialized, and the GVF snakes are then evolved automatically to detect floe boundaries and separate the connected floes. The GVF snake-based method is applicable to identify non-ridged ice floes, especially in the marginal ice zone (MIZ) and managed ice resulting from offshore operations in sea ice. It will generally experience less over- or under-segmentation than the watershed method.

Because some holes and smaller ice pieces may be contained inside larger floes by using the GVF snake-based segmentation method, all the segmented ice floes are arranged in the order of increasing size. The morphological cleaning and hole filling algorithms are then performed to the arranged ice floes in sequence to enhance their shapes. This results in the final identification of the individual ice floes in the image. Based on the identification result, different types of sea ice pieces are distinguished, and the image is divided into four layers: ice floes, brash ice, slush, and water. This then makes it possible to present a color map of the ice floes and brash pieces based on their sizes. It also makes it possible to present the corresponding ice floe size distribution histogram and to store all geometric data for each individual ice floe.

The introduced sea ice image processing algorithms have been developed and applied in Arctic offshore engineering for calculating the ice concentration and analyzing the floe size distribution from full-scale sea ice images. They have also been applied to images from dynamic positioning (DP) experiments in the sea ice model basin at the Hamburg Ship Model Basin (HSVA) for determining the ice concentration and floe size distribution from images and for identifying individual ice floe sizes from surveillance videos. Particularly, in order to obtain the floe size distributions from the model basin images, in which the ice floes were modeled deterministically as rectangular shapes with predefined side lengths. After the GVF snake had successfully identified the individual ice floes, an additional step was employed to fit the identified floes to rectangular shapes and from that identify overlaps between floes.

The MATLAB® source codes of the algorithms for the image processing methods discussed in the book are available as online material. We hope that the readers of the book will benefit from it and find the text and supplemental material useful in their work.

MATLAB$^{®}$ is a registered trademark of The MathWorks, Inc. For product information please contact:
The MathWorks, Inc.
3 Apple Hill Drive
Natick, MA, 01760-2098 USA
Tel: 508-647-7000
Fax: 508-647-7001
E-mail: info@mathworks.com
Web: www.mathworks.com

Acknowledgments

We would like to acknowledge Professor Chi-Hau Chen for his kind invitation to write this book and his follow-up of the process afterwards.

The research leading to this book is based on the PhD work "Image Processing for Ice Parameter Identification in Ice Management" by the 1st author. This research was supported by the Research Council of Norway (RCN), partly through the KMB Arctic DP project, RCN no. 199567, with supporting industrial partners Statoil, Kongsberg Maritime, and DNV GL, and partly through the Centre for Research-based Innovation on Sustainable Arctic Marine and Coastal Technology (CRI SAM-CoT), RCN project no. 203471. We would therefore like to thank the KMB Arctic DP project and the CRI SAMCoT research center, and all of their collaborating partners, as the arenas where the research leading to this book has been done. We also would like to thank the MARTEC ERA-NET project DYPIC: Dynamic Positioning in Ice Covered Waters (RCN project no. 196897) and the Hamburg Ship Model Basin for providing experimental data.

We would like to express our thanks to Professor Sveinung Løset, Director of CRI SAMCoT, for all his support during this research work, as well as Dr. Takenobu Toyota, Dr. Francesco Scibilia, and Professor Lars S. Imsland for their constructive feedback during the PhD defense of the 1st author.

We would like to thank Dr. Wenjun Lu for implementing the image processing algorithms to extract ice parameters for real Arctic offshore engineering problems, to the Swedish Polar Research Secretariat and the crew of icebreakers Oden and Frej during the Oden Arctic Technology Research Cruise 2015, and also all our other colleagues and partners from the KMB Arctic DP project, the CRI SAMCoT research center, and the Department of Marine Technology at the Norwegian University of Science and Technology (NTNU) for their help and cooperation.

As 1st author of this book, I would like to express my sincere gratitude to my PhD supervisor, Prof. Roger Skjetne, who is also the 2nd author of this book. Without him, I would not have had the opportunity to work on this specific topic, which led to the present book. Moreover, I am deeply grateful to my family. I am greatly indebted to my mother, Yuzhi Chen, my father, Zhijian Zhang, my sister, Yu Zhang, and my little nephew, Minxuan Xiao, for their support and encouragement throughout the preparation of this book. I appreciate my mother-in-law, Yulan Ge, for assisting me in taking care of my newborn son for a few months so that I could get more time to write the book. My special appreciations go to my husband, Dr. Biao Su, for his love, confidence, and support in motivating me to complete this book, and my 1-year-old son, Weiyang Su, for bringing me a great deal of enjoyment during this period of time.

Trondheim, Norway

Qin Zhang

As 2nd author, I am thankful to my family for being patient with me. But most of all I am grateful to 1st author Dr. Qin Zhang, who has done the major part in writing this book. I know it is not always easy to collaborate with a professor drowning in university duties.

In addition to the abovementioned projects, I would like to acknowledge the Centre for Autonomous Marine Operations and Systems (NTNU AMOS), RCN project no. 223254. I would also like to thank the Center for Control, Dynamical Systems, and Computation at the University of California at Santa Barbara, and especially Professor Andrew R. Teel, for inviting me for a research sabbatical 2017-2018 where I have managed to complete my part of this book project.

Santa Barbara, CA, USA. Roger Skjetne

Contributors

Qin Zhang

Department of Marine Technology, Norwegian University of Science and Technology (NTNU), Trondheim, Norway

Dr. Qin Zhang received her MSc degree in signal and information processing from the East China University of Science and Technology, Shanghai, China, in 2010 and PhD degree in 2015 at Norwegian University of Science and Technology (NTNU), Trondheim, Norway. She was awarded Chinese government award for outstanding student abroad in 2013. Qin Zhang became a researcher in 2015 at the Centre for Research-based Innovation (CRI) Sustainable Arctic Marine and Coastal Technology (SAMCoT), NTNU. Her research interests include remote sensing, and image and sensory data processing.

Roger Skjetne

Department of Marine Technology, Norwegian University of Science and Technology (NTNU), Trondheim, Norway

Prof. Roger Skjetne received his MSc degree in 2000 from the University of California at Santa Barbara (UCSB) and his PhD degree in 2005 from the Norwegian University of Science and Technology (NTNU) on control engineering. He holds an Exxon Mobil prize for his PhD thesis at NTNU. Prior to his studies, he worked as a certified electrician for Aker Elektro AS on numerous oil installations for the North Sea, and in 2004-2009 he was employed in Marine Cybernetics AS, working on Hardware-in-the-Loop simulation for testing marine control systems. From August 2009 he has held the Kongsberg Maritime chair of Professor in Marine Control Engineering at the Department of Marine Technology at NTNU. His research interests are within dynamic positioning of marine vessels, Arctic stationkeeping and Ice Management systems, control of shipboard hybrid electric power systems, nonlinear motion control of marine vehicles, and autonomous marine robots. Roger Skjetne is leader of the ice management work package in the CRI Sustainable Arctic Marine and Coastal Technology (SAMCoT), associated researcher in the CoE Centre for Ships and Ocean Structures (CeSOS) and CoE Autonomous Marine Operations and Systems (NTNU AMOS), principal researcher in the CRI on Marine Operations (MOVE), and he was project manager of the KMB Arctic DP research project. In 2017-2018, he was a visiting research scholar at the Center for Control, Dynamical-systems and Computation at the University of California at Santa Barbara. He is also co-founder of the two companies BluEye Robotics AS and Arctic Integrated Solutions AS.

List of Figures

List of Tables

1 Introduction

1.1 RESEARCH BACKGROUND

Sea ice, defined as any form of ice that forms as a result of sea water freezing [91], occurs primarily in the polar regions and covers approximately 7% of the total area of the world's oceans [168]. Sea ice is turbulent because of wind, wave, and temperature fluctuations, and it influences the movement of ocean waters, fluxes of heat, and circulation between atmosphere and ocean [1]. Sea ice plays important roles in climatology, meteorology, oceanography, physics, maritime navigation, marine biology, Arctic (and Antarctic) offshore operations, and world trade [137]. For example, if gradual warming melts sea ice over time, the abnormal changes in the amount of sea ice can affect the habitats of the animals that live in the polar regions, and it can disrupt normal atmosphere/ice/ocean momentum transfer and heat exchange that thereby may lead to further changes in global climate [2]. Moreover, the prevalence of sea ice will be a determining factor to human activities in the Arctic regions, such as scientific voyages, oil and gas activities, and Arctic shipping through the availability of the Arctic sailing routes from northern European to northern Pacific ports.

Ice concentration, ice floe size distribution, and ice types are important parameters in the field observations of sea ice. Because the sizes of the ice floes and brash ice can range from about one meter to a few kilometers, the temporally and spatially continuous field observations of sea ice are necessary for safe marine activities and understanding of the Arctic climate change. To that end, one of the most efficient ways to observe the ice conditions in the oceans is by using satellite, aerial, or nautical imagery and applying digital image processing techniques to the ice image data.

The analysis of image information obtained from remote sensors can reduce or suppress the ambiguities, incompleteness, uncertainties, and errors of the object and the environment via various processing techniques. It can also make the information of the object and environment more accurate and reliable by maximizing the use of image information from a variety of information sources, obtaining a more comprehensive and distinct environment. Therefore, various types of remotely sensed data and imaging technologies have been aiding the development of sea ice observation. Particularly the satellite observing systems and corresponding data processing algorithms have been widely used in the determination of sea ice parameters, such as extracting ice concentration [141, 34, 136, 79], classifying ice types [59, 26, 144, 14, 180, 47], and analyzing ice floe properties [7, 86, 145]. Nowadays the ice concentration data on a global scale has become available on a daily basis due to the development of microwave satellite sensors. According to this innovation, it has become possible to monitor the variability of sea ice extent on a global basis. However, it is still a big issue to predict the sea ice behavior in the numerical sea ice model due to the lack of our knowledge about the sub-grid scale information of

1

sea ice. To solve this, the attention to ice floe size distribution has recently been increasing. Not limited to the climate problem, sea ice information on a sub-grid scale is also an important issue to the maintenance of the infrastructure built in the polar ocean.

Focusing on a relatively small scale, the use of cameras as sensors on mobile sensor platforms (e.g., aircrafts, marine vessels, or other unmanned vehicles) can be explored for ice motion monitoring and characterization of ice conditions. Cameras as sensors have the potential of continuous measurements with high precision that allows capturing a wide range of the sea ice field, from a few meters to hundreds of meters. The visible image data obtained from cameras have high resolution, which is particularly important for providing detailed localized information of sea ice to collect observational data on an operational basis [58, 69]. The information of the object and environment provided by such visible images are close to human visual perception on them both in tonal structure and resolution, meaning that determination of sea ice characteristics via visible sea ice images is similar to manual visual observation. Thus cameras can be used as supplementary means to necessary and important information of the actual ice conditions for validation of theories and estimation of parameters in combination with other instruments for sea ice remote sensing [70].

Due to those advantages, visible camera imagery becomes one of the most informative remote sensing tools and has been applied to sea ice observation [80, 115, 149, 54]. Rothrock and Thorndike [129] analyzed the mosaic of aerial photographs of summer pack ice to measure the sea ice floe size distribution. Tschudi et al. [161] used a downward-looking video camera mounted on the underside of an aircraft to acquire images of the ice surface for the determination of the areal coverage of melt ponds, new ice, and open water. Lu et al. [93] extracted geometric parameters of sea ice floes in the marginal ice zone (MIZ) of Prydz Bay from the aerial sea ice photographs acquired by a helicopter-mounted camera during the 21st Chinese National Antarctic Research Expedition from December 2004 to February 2005. In order to compare the ground truth data with satellite data, Muramoto et al. [110, 109] measured sea ice concentration and floe size distribution continuously along the ship's route from ship mounted video images photographed in ice-covered water in the Southern Ocean between Fremantle, Australia and Syowa Station in 1988; while Hall et al. [54] obtained systematic sea ice concentrations from a ship-borne camera system during the R.V. Jan Mayen scientific cruise to the Greenland Sea, March 2000, and compared the results with SSM/I (Special Sensor Microwave/Imager) -derived sea ice concentrations along the cruise track. For a similar ship-based ice condition imagery acquisition, Weissling et al. [169] developed a camera system together with image processing techniques to derive ice concentration, ice types, floe sizes, and area of deformed ice during the 2007 Sea Ice Mass Balance in Antarctic (SIMBA) cruise, while Lu and Li [92] conducted a study on estimating the geometric distortion induced by oblique photography based on the photogrammetric theory and developing an accurate algorithm to obtain the ice concentration and floe size from a shipborne camera. A sea ice digital image collection and processing system was utilized by Ji et al. [69] to monitor the sea ice parameters in the JZ20-2 oil-gas field of the Liaodong Bay, and by this system, ice thickness, ice concentration, and ice velocity of the whole ice period in the Bohai Sea were determined continuously during the winter of 2009-2010.

Despite the advantages of using visible cameras for sea ice observation, an important requirement is clear weather and sight when capturing sea ice image data. Moreover, one of the major problems of surveying sea ice via cameras has been the difficulty in image processing for numerical extraction of sea ice information, which is vital for feeding relevant parameters into models, estimating the sea ice properties, and understanding the behavior of sea ice, especially on a relatively small scale. Since sea ice is a complicated multi-domain process, it is not easy to analyze the sea ice images quantitatively. The lack of a systematic and standard method for image processing has hampered the understanding of the dynamical properties of sea ice on small scale.

Therefore, the purpose of this book is to introduce image processing algorithms to extract useful ice information (particularly the information of ice floe boundaries) from small scale visible ice images in a systematic and objective way. The introduced ice image processing algorithms may potentially help to illuminate the momentum exchange from atmosphere to ice [148], estimate the heat fluxes in the ice-covered regions [104] and the melting rate of ice floes [147], model the rheology of ice for different seasons, thickness, types, and regions [18], and possibly provide a clue to the understanding of ice-floe formation processes [155]. These algorithms can also be used further to develop tools, based on the processed ice data, that can supplement data provided by other sensors onboard the ship or a buoy in form of ice concentration, ice floe boundaries, and ice types in the surrounding region, and be applied for decision support in Arctic offshore operations [139, 164, 108]. Furthermore, the image processing methods described in this book may potentially be applied further to other research fields, such as biomedicine, material technology, and other observation/engineering areas.

1.2 SEA ICE PARAMETERS

The ice parameters considered in this book include ice concentration, ice types, ice floe positions and areas, and floe size distribution. Corresponding definitions for these parameters in this book are specified in the following.

1.2.1 ICE CONCENTRATION

Ice concentration (IC) is the ratio of ice on unit area of sea surface. To obtain IC from a visual ice image, only the visible ice can be considered, including brash ice and, if visible in the image, submerged ice. With the image area, the height of the image taken above the ice sheet, and the segmentation, which is the identification of the ice pixels from the water pixels, the actual area of sea ice and sea surface can be derived. However, the actual domain area is not necessary for calculating the ice concentration.

In simplified terms, ice concentration from a digital visible image is, in this book, defined as the area covered by visible ice observable in the 2-dimensional image, taken vertically from above, over the total sea surface domain area of the image.

A digital image is a numerical representation of a 2-dimensional picture as a finite set of values called pixels. Hence, ice concentration can be derived by calculating the fraction of the number of pixels of visible ice to the total number of pixels within the image domain. An image may contain parts of land or other non-relevant areas. Thus, the domain area is an effective area within the image after the non-relevant parts have been removed. The ice concentration is then given by:

$$
\begin{aligned}
IC &= f(image\ area,\ height\ above\ ice\ sheet,\ segmentation) \\
&= \frac{Area\ of\ all\ visible\ ice\ within\ domain}{Actual\ domain\ area} \\
&= \frac{Number\ of\ pixels\ of\ visible\ ice\ in\ the\ image\ domain}{Total\ number\ of\ pixels\ in\ the\ image\ domain}
\end{aligned}
\tag{1.1}
$$

1.2.2   ICE TYPES

Various types of sea ice can be found in ice-covered regions, and different types of sea ice have different physical properties. As defined in Løset et al. [91]:

- Floe is any relatively flat piece of sea ice 20 m or more across. It is subdivided according to horizontal extent. A giant flow is over 10 km across; a vast floe is 2 to 10 km across; a big floe is 500 to 2000 m across; a medium floe is 100 to 500 m across; and a small floe is 20 to 100 m across.
- Ice cake is any relatively flat piece of sea ice less than 20 m across.
- Brash ice is accumulations of floating ice made up of fragments not more than 2 m across and the wreckage of other forms of ice. It is common between colliding floes or in regions where pressure ridges have collapsed.
- Slush is snow that is saturated and mixed with water on land or ice surfaces, or as a viscous floating mass in water after heavy snowfall.

In this book, for simplicity, the size of the sea ice piece is the only criterion to distinguish ice floe and brash ice. That is, any relatively flat piece of sea ice 2 m or more across is considered as "ice floe", while any relatively flat piece of sea ice less than 2 m across is considered as "brash ice (piece)". The residual of ice pixels are considered as "slush".

1.2.3   ICE FLOE SIZE AND FLOE SIZE DISTRIBUTION

The estimation of ice floe size and floe size distribution (FSD) among the "ice floes" gives an important set of parameters from ice images. In image processing, the ice floe size can be determined by the number of pixels in the identified floe. If the focal length f and camera height are available, the actual size in SI unit of the ice floes and floe size distribution can also be calculated by converting the image pixel size to its SI unit size [92].

In practice, other parameters are typically used to represent the size of the floes, such as the "representative diameter". The algorithms proposed in this book produce an ice image data structure (as seen in Appendix B) containing a complete database of all floes and brash ice in the image, where the pixels of each floe is stored. Hence, any 2-dimensional geometric parameter can easily be calculated from the database. The floe size distribution can thus be easily recalculated based on the "representative diameter" of the floes.

1.3 APPLICATIONS OF DIGITAL IMAGE PROCESSING TECHNIQUES FOR ICE PARAMETER IDENTIFICATION

Digital images were first used for transferring newspaper pictures between London and New York in the early 1920s, where the pictures were coded for the submarine cable transmission and reconstructed by a special telegraph printer at the receiving end. The concept of digital image processing became meaningful and many of the digital image processing capabilities were developed in the 1960s when both hardware and software of computer technology were developed powerful enough to carry out image processing algorithms. In the 1970s, digital image processing techniques began to be used in the space program, medical imaging, remote sensing, and astronomy as cheaper and dedicated computer hardware became available. Until now, with the rapid development of computer technology, the use of digital image processing techniques has been growing by leaps and bounds, and has achieved success in many applications such as remote sensing, industrial inspection, medicine, biology, astronomy, law enforcement, defense, etc. [48].

In most cases, human manual interpretation is simply impossible, and the only feasible solution for information extraction from images is through digital image processing by a computer. Digital image processing algorithms, implemented by computers, are important to replace humans in the interpretation of image data. Many image processing algorithms have been developed for the analysis of sea ice statistics and ice properties from remotely sensed sea ice images, and in this section we will give an overview of some of the relevant literature in this field.

1.3.1 ICE CONCENTRATION CALCULATION

From Equation 1.1, it is clear that the estimation of ice concentration by using ice imagery data is equivalent to the discrimination of ice pixels from water pixels. Due to the fact that ice is normally brighter than water, a thresholding approach is typically used for extracting ice from water pixels [54, 169, 185]. For instance, Markus and Dokken [103] propose that sea ice pixels can be determined by adapting thresholds between ice and open water based on local intensity distributions, while Johannessen et al. [70] introduces an algorithm of sea ice concentration retrieval from ERS (European Remote Sensing) SAR (Synthetic Aperture Radar) images by using two thresholds to separate open water from thick ice.

Ice concentration derivation is usually associated with ice type classification, since all types of sea ice should be taken into account for calculating ice concentration. Hence, the algorithms for classifying ice types, such as unsupervised and supervised classification [169], texture features [89], and neural networks [77], etc., can also be used for calculating ice concentration. The ice concentration is then derived by summing up the concentrations of multiple ice types existing in the ice image.

1.3.2 SEA ICE TYPE CLASSIFICATION

Unsupervised and supervised classification algorithms are popular for sea ice type classification [81, 55, 42, 143, 133, 112, 138, 43, 142, 181]. In an unsupervised classification approach, pixels are assigned to classes based on their spectral properties, without the user having any prior knowledge of the existence of those classes; while in a supervised classification approach, pixels are grouped based on the knowledge of the user by providing sample classes to train the classifier [71]. Hughes [64] examined the use of an unsupervised k-means clustering method for automatic classification of the data from 7 SSM/I channels, and he demonstrated that it is possible to obtain classifications of the different ice regimes both in the seasonal and perennial ice cover by clustering using emissivities from all channels. Dabboor and Shokr [37] proposed an iteratively supervised classification approach that utilized a complex Wishart distribution-based likelihood ratio (LR) and a spatial context criterion to discriminate sea ice types for polarimetric SAR data.

Image features, particularly texture features that characterize local and statistical properties of regions in an image, have been widely used in the classification of sea ice types [63, 138, 31, 142, 26, 27]. Several research works have been done on gray-level co-occurrence matrices (GLCM) texture analysis [56] for sea ice image classification [138, 101, 89]. Many important parameters need to be defined for GLCM. Soh and Tsatsoulis [142] quantitatively evaluated GLCM texture parameters and representations, and they determined best textural parameters and representations for mapping texture features of SAR sea ice imagery. They also developed three GLCM implementations and evaluated these developed implementations by a supervised Bayesian classifier on sea ice textural contexts. Other texture analysis methods, such as Gabor, and Markov random fields (MRF), can also used in sea ice image classification. Clausi [25] compared the ability of texture features based on GLCM, Gabor, MRF, and the combination of these three methods for classifying SAR sea ice image.

Neural networks have also been applied to classifying sea ice types [74, 14, 181]. For examples, Comiso [33] utilized a back-propagation neural network to improve the classification by using the unsupervised ISODATA cluster analysis results to train the system. Hara et al. [55] developed a neural network that employed the learning vector quantization (LVQ) method to perform the initial clustering and improved the results by an iterative maximum likelihood (ML) method for the classification of sea ice in SAR imagery. Pedersen et al. [119] used a feed-forward back propagation neural network with 3 layers for sea ice type classification based on texture features.

Besides the classification methods mentioned above, Yu and Clausi [180] developed a so-called iterative region growing using semantics (IRGS) algorithm that combined image segmentation and classification for classifying the operational SAR sea ice imagery. In this IRGS algorithm, the watershed algorithm [167] was first used to segment the image into small homogeneous regions, then the MRF-based labeling and the region merging processes were performed iteratively until the merging cannot be performed further. The IRGS algorithm has been applied to polygons from sea ice maps provided by the Canadian Ice Service (CIS) for classifying sea ice types [112], and further extended for polarimetric SAR image classification by incorporating a polarimetric feature model based on the Wishart distribution and modifying key steps [179].

It should be noted that the works mentioned above mainly classified sea ice into first-year, multi-year, and young ice. Those sea ice types are different from the ice types that we classify in this book, as described in Section 1.2.2.

1.3.3 ICE FLOE IDENTIFICATION

Ice floe boundary detection is crucial to the identification of individual floes, particularly those that are connected to each other, and to further determination of floe size distribution and floe shape identification. Zhang et al. [185] applied and compared traditional derivative and morphology edge detection algorithms for the identification of floe boundaries. These edge detection algorithms, however, cannot easily detect the boundary between connected floes. This issue challenges the boundary detection of individual ice floes and significantly affects ice floe size analysis. Therefore, the boundary hidden by an apparent connection between ice floes should be identified.

To mitigate this issue, many efforts have been made. Toyota et al. [157, 156] separated closely distributed ice floes by setting a threshold higher than the ice-water segmentation threshold, while Lu et al. [93] used a local dynamic thresholding approach to segment ice floes. However, these thresholds still did not work well when the ice floes were connected. Consequently, they manually separated the connected ice floes. Banfield [6] and Raftery [7] introduced a mathematical morphology together with the principal curve clustering to identify ice floes and their boundaries in a near fully automated manner. In their approach, the image was firstly binarized using the thresholding method. The erosion-propagation (EP) algorithm was then used to provide a preliminary clustering of the boundary pixels and to produce a collection of objects as floe candidates. To remove subdivisions caused by the EP algorithm, they developed a method based on an algorithm for clustering about closed principal curves to determine which floes should be merged. Soh et al. [145] introduced the restricted growing concept and used probabilistic labeling to extract individual ice floes with minimal human intervention requirement. Their algorithm first shrank the floes to their core floes (the floes whose original sizes were reduced) by using the probabilistic labeling for creating the separation among the connected floes. Then the algorithm grew the floes back to their original sizes and shapes restricted by the boundaries of their corresponding skin floes (the floes whose original sizes and shapes have been preserved; they are usually interconnected and could encompass one or more core floes) while preserving the existing separation among the core floes. Blunt et al. [13] and Zhang et al. [187] adopted the watershed transform to separate connected ice floes. Since the watershed segmentation has a tendency to over-segment, a further correction step is needed. Blunt et al. [13] manually removed these over-segmentations. Zhang et al. [187], on the other hand, automatically removed the over-segmented lines that had no concave ending point between the corresponding segmented ice floes by the watershed algorithm.

The connected ice floe separation methods mentioned above require human interventions, or operate on the binarized sea ice images where the actual information of floe boundaries are lost. Thus, these methods will challenge the identification of individual ice floes from the ice images where mass of ice floes connect together, particularly from MIZ images. To overcome this, Zhang and Skjetne [184, 183] adopted a gradient vector flow (GVF) snake algorithm to identify the floe boundaries, where the distance transform was used to automatically initialize the contours for the GVF snake algorithm.

1.4 BOOK SCOPE AND STRUCTURE

Focusing on the image processing algorithms for obtaining information regarding small scale features from visible ice images, the scope of this book includes the following:

- Segmenting ice regions from water regions.
- Detecting ice floe boundaries and locating floe positions.
- Classifying different types of sea ice.
- Estimating ice concentration and floe size distribution.

A brief introduction concerning background, motivation, and objectives of this book is given in Chapter 1. The definitions of ice concentration from a digital visible ice image, sea ice types (i.e., ice floe, brash ice, and slush), and ice floe size distribution are also specified in Chapter 1.

Before starting with various sea ice image processing algorithms, some basic concepts and methodologies relevant to digital image processing, such as image types, histogram, convolution, set and logical operations, image interpolation, and so on, are introduced in Chapter 2. This will help readers without background on digital image processing to go through the book and understand the subject and thereby appreciate the use of digital image processing in sea ice parameter identification problems.

For the purpose of determining the ice concentration from ice images, Chapter 3 presents the Otsu thresholding and k-means clustering methods to detect ice pixels, since the key for deriving ice concentration via digital image processing techniques is to distinguish the ice pixels from water pixels in ice images.

Automatic identification of individual floe boundaries is crucial in obtaining the size, shape, and location of ice floes and further determination of floe size distribution from ice images. Chapter 4 shows how to apply derivative and morphology edge detection algorithms to sea ice images. These methods are good when the floes are clearly distinguishable. However, derivative edge detection is sensitive to weak boundaries and noise, and it often produces non-closed boundaries, meaning that junctions between ice floes may be difficult to identify when the floes are seemingly connected in the ice images. In contrast, morphology edge detection gives a good description of the object shape, generates closed boundaries, and may be used to separate weakly connected ice floes; some boundary information may still be lost, however, and strongly connected ice floes cannot be separated.

Because of the inability of both derivative and morphology edge detection methods for separating the connected ice floes in the ice images, Chapter 5 shows how the watershed transform can be used to segment such connected sea ice floes. In this approach, it is assumed that each ice floe has a convex boundary and that the junction line between two connected ice floes has at least one concave ending point. All the ice floes in the image are then first identified by the Otsu thresholding or k-means clustering algorithm, converting the gray-scale image into a binary image. Then the seed points for the individual ice floes are located, and the watershed algorithm is carried out to segment the image. After the watershed transform, the convexity of each pair of ending points is checked. Two neighboring floes whose junction line ending points are both convex are automatically merged to remove over-segmentation.

The watershed-based segmentation method introduced in Chapter 5 operates on the binary images, focusing on the morphological characteristics of ice floes rather than on the real boundaries. To utilize the real boundary information of the ice floes, the so-called gradient vector flow (GVF) snake algorithm is presented in Chapter 6 to separate seemingly connected floes into individual floes due to its capability of detecting weak edges. To avoid user interaction and to reduce the time requirement for running the GVF snake algorithm, an automatic contour initialization method is further proposed based on the distance transform.

After ice floe segmentation, some holes or smaller ice pieces may be contained inside larger ice floes. An ice shape enhancement algorithm based on morphological cleaning is therefore introduced in Chapter 7. The objective here is to enhance the ice floe shapes and accomplish the identification of individual ice floes together with other ice types. Once the individual ice floes are identified, the floe size distribution can be calculated from the resulting data. Moreover, two case studies are also given in Chapter 7 to discuss the usage of the proposed sea ice image processing algorithms.

The proposed sea ice image processing algorithms have been developed and applied in the Arctic DP project (Dynamic Positioning of vessels in Arctic drifting sea ice conditions), which was a research project for safe and green dynamic positioning operations of offshore vessels in an Arctic environment [139, 183]. It has further been applied in the research center CRI Sustainable Arctic Marine and Coastal Technology (SAMCoT) related to images from offshore ice management operations in an interior ice zone [94]. Chapter 8 presents the applications of the developed algorithms for the analysis of sea ice images with Arctic offshore engineering.

Not limited to the sea ice images, the proposed ice image algorithms have also been developed for the indoor model sea ice images and surveillance videos from the dynamic positioning (DP) experiments in sea ice model basin at the Hamburg Ship

Model Basin (HSVA). The applicability of the proposed sea ice image processing algorithms to ice management in model-scale is illustrated in Chapter 9 for further use in ice-load numerical simulations and dimensioning studies.

Because of the oblique angles or optical lenses of the cameras, the captured images are usually distorted in practice. Such geometric distortions must be removed, otherwise they will affect the generated statistics of the ice floes analysis. Perspective and radial distortions are the two most common geometric distortions in sea ice images. Thus, both analytical and simplified calibration methods for correcting perspective and radial distortions are given in Appendix A.

Finally, the sea ice image algorithms proposed in this book are used to populate a database of all floes and brash ice pieces in the image and their properties. The details of this ice image data structure, based on MATLAB®, are presented in Appendix B for reference.

2 Digital Image Processing Preliminaries

A digital image in a 2-dimensional discrete space is the sampling and quantization of a 2-dimensional continuous space, being a projection of a picture of objects and background in 3-dimensional space. A digital image is composed of 2-dimensional array elements arranged in rows and columns. Those elements are the so-called pixels, and each of them holds a particular value to represent the picture at its location.

For a mathematical expression, a digital image can be represented as a function $f(x,y)$, where (x,y) are integers and f is a mapping that assigns an intensity value to each distinct pair of coordinates (x,y). A digital image with M rows and N columns, which we say that the image is of size $M \times N$, can also be represented as a matrix:

$$f = \begin{bmatrix} f(1,1) & f(1,2) & \cdots & f(1,y) & \cdots & f(1,N) \\ f(2,1) & f(2,2) & \cdots & f(2,y) & \cdots & f(2,N) \\ \vdots & \vdots & \ddots & \vdots & \ddots & \vdots \\ f(x,1) & f(x,2) & \cdots & f(x,y) & \cdots & f(x,N) \\ \vdots & \vdots & \ddots & \vdots & \ddots & \vdots \\ f(M,1) & f(M,2) & \cdots & f(M,y) & \cdots & f(M,N) \end{bmatrix} \quad (2.1)$$

where $f(x,y)$ $(1 \leq x \leq M, 1 \leq y \leq N)$ is the finite and quantized value that represent the gray scale or color of the image at the point (x,y).

In this chapter, some beforehand knowledge about digital image processing relevant to the sea ice image processing algorithms presented in this book is introduced.

2.1 IMAGE TYPES

Each of the pixels in a digital image has a pixel value that represent the brightness or color it should be. Based on this, there are two important types of digital images: grayscale images and color images.

2.1.1 GRAYSCALE IMAGE

A grayscale image is a monochrome image that is made up of pixels in different shades of gray from black to white. The pixel value is a single number corresponding to the grayscale of the image at a particular location. An example of a grayscale image is shown in Figure 2.1.

The number of gray scales is typically an integer power of 2. That is, the grayscale intensity is stored as a B-bit integer giving 2^B possible different shades of gray in the range $[0, 2^B - 1]$, where B is the number of intensity bits. When $B = 1$ in particular,

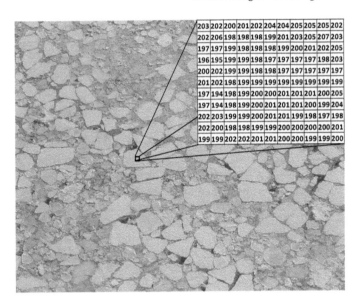

Figure 2.1 Pixel values in a grayscale image.

the image is a binary image. Since the pixel value in a binary image is a 1-bit number, a binary image is a logical array of 0s and 1s, and it has only two colors, usually black or white as seen in Figure 2.2.

1	1	1	1	1	1	1	1	1	1	1	1	1	1	1	1
1	1	1	1	1	1	1	1	1	1	1	1	1	1	1	1
1	1	0	0	0	0	0	0	0	0	0	0	0	0	1	1
1	1	0	0	0	0	0	0	0	0	0	0	0	0	1	1
1	1	0	0	1	1	1	1	1	1	1	1	0	0	1	1
1	1	0	0	1	1	1	1	1	1	1	1	0	0	1	1
1	1	0	0	1	1	0	0	0	0	1	1	0	0	1	1
1	1	0	0	1	1	0	0	0	0	1	1	0	0	1	1
1	1	0	0	1	1	0	0	0	0	1	1	0	0	1	1
1	1	0	0	1	1	0	0	0	0	1	1	0	0	1	1
1	1	0	0	1	1	1	1	1	1	1	1	0	0	1	1
1	1	0	0	1	1	1	1	1	1	1	1	0	0	1	1
1	1	0	0	0	0	0	0	0	0	0	0	0	0	1	1
1	1	0	0	0	0	0	0	0	0	0	0	0	0	1	1
1	1	1	1	1	1	1	1	1	1	1	1	1	1	1	1
1	1	1	1	1	1	1	1	1	1	1	1	1	1	1	1

Figure 2.2 Pixel values in a binary image.

2.1.2　COLOR IMAGE

2.1.2.1　The RGB color space

A color image is formed by a combination of individual 2-dimensional images. For example, in the RGB (red, green, and blue) color system, any color can be created by a weighted combination of the three primary colors red, green, and blue. Hence, a color image can be viewed as a combination of red, green, and blue individual component images. As seen in Figure 2.3, an RGB image is an $M \times N \times 3$ array of color pixels. Any pixel C located at (x,y) in the RGB image holds three numbers $R(x,y)$, $G(x,y)$ and $B(x,y)$ corresponding to the red, green, and blue values of the image at a point (x,y):

$$C(x,y) = \begin{bmatrix} R(x,y) \\ G(x,y) \\ B(x,y) \end{bmatrix} \tag{2.2}$$

For this reason, many of the techniques developed for grayscale images can be extended to color images by processing the three component images individually.

2.1.2.2　The CMY and CMYK color spaces

The CMY (cyan, magenta, and yellow) color model is a subtractive color representation. It is typically used in color printing because cyan, magenta, and yellow are the primary colors of pigments. The CMY color model can be transformed from the RGB model by:

$$\begin{bmatrix} C \\ M \\ Y \end{bmatrix} = \begin{bmatrix} 1 \\ 1 \\ 1 \end{bmatrix} - \begin{bmatrix} R \\ G \\ B \end{bmatrix} \tag{2.3}$$

where the tristimulus values in the RGB color model are normalized to the range $[0,1]$. Figure 2.4 presents the CMY components of the color image shown in Figure 2.3.

In practice, to produce true black color for printing without using excessive amounts of CMY pigments, black, called the key (K), is added as a fourth color, giving rise to the CMYK color model. The conversion between the CMYK and RGB is given by [121]:

$$\begin{bmatrix} C \\ M \\ Y \\ K \end{bmatrix} = \begin{bmatrix} 1 \\ 1 \\ 1 \\ 0 \end{bmatrix} - \begin{bmatrix} R \\ G \\ B \\ 0 \end{bmatrix} - K_b \begin{bmatrix} u \\ u \\ u \\ -b \end{bmatrix} \tag{2.4}$$

where

$$K_b = \min\{1-R, 1-G, 1-B\} \tag{2.5}$$

and u $(0 \leq u \leq 1)$ is the under-color removal factor, and b $(0 \leq b \leq 1)$ is the darkness factor.

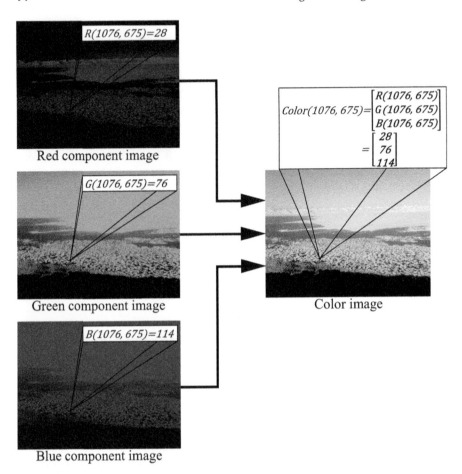

Figure 2.3 Pixel values in a RGB image.

(a) Cyan component image.   (b) Magenta component image.   (c) Yellow component image.

Figure 2.4 CMY components of the color image shown in Figure 2.3.

2.1.2.3   The HSI color space

Alternative to the RGB, CMY and CMYK color spaces, a hue-saturation color coding method, HSI (hue, saturation, and intensity), is also commonly used, particularly in the image processing algorithms based on color descriptions. Hue is an attribute that describes a pure color, while saturation (purity) is a measure of the degree to which pure color is diluted by white light. The HSI color model decouples the intensity component from the hue and saturation in a color image [49], and it can be obtained from the RGB color model by [121]:

$$
\begin{bmatrix} I \\ V_1 \\ V_2 \end{bmatrix} = \begin{bmatrix} \frac{1}{3} & \frac{1}{3} & \frac{1}{3} \\ \frac{-1}{\sqrt{6}} & \frac{-1}{\sqrt{6}} & \frac{2}{\sqrt{6}} \\ \frac{1}{\sqrt{6}} & \frac{-1}{\sqrt{6}} & 0 \end{bmatrix} \begin{bmatrix} R \\ G \\ B \end{bmatrix} \tag{2.6a}
$$

$$
H = \arctan\left(\frac{V_2}{V_1}\right) \tag{2.6b}
$$

$$
S = \sqrt{V_1^2 + V_2^2} \tag{2.6c}
$$

Figure 2.5 presents the HSI components of the color image shown in Figure 2.3.

(a) Hue component image.   (b) Saturation component image.   (c) Intensity component image.

Figure 2.5   HSI components of the color image shown in Figure 2.3.

2.1.3   INDEXED IMAGE

An indexed image is an image where pixel values are integer numbers and those numbers are used as pointers (indices) to color values stored in the colormap or "palette". An indexed image consists of two components: an image matrix of integers and a colormap. The colormap is an $m \times 3$ matrix of class double where m is the number of distinct image colors in the colormap. Each row of the colormap matrix specifies the red, green, and blue values ranging from 0 to 1 for a single color. Each pixel value of the integer matrix points to which of the colors in the colormap that applies to that pixel. This reduces the amount of data that needs to be stored with the image. Figure 2.6 gives an example to illustrate the structure of an indexed image.

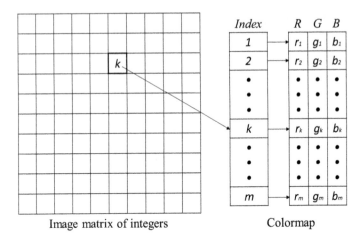

Image matrix of integers · Colormap

Figure 2.6 Structure of an indexed image.

2.2 IMAGE HISTOGRAM

The histogram of an image is a statistic showing the distribution of the pixel intensity values. For an image with L possible intensity levels in the range of $[0, L-1]$, the histogram is the number of pixels in the image at each different intensity level, defined as the discrete function:

$$h(r_k) = n_k \qquad (2.7)$$

where r_k is the kth intensity level in the interval $[0, L-1]$, and n_k is the number of pixels in the image whose intensity level is r_k. Note that $L = 2^B$ where B is the bit depth of the image.

For a grayscale image that has L different possible intensities, L numbers will be displayed in its histogram to show the distribution of pixels among those grayscale values. An example of the histogram of an 8-bit grayscale image, which has 256 possible intensity levels, is shown in Figure 2.7. For a color image, three individual histograms of red, green, and blue channels can be taken, as shown in Figure 2.8.

A histogram is usually normalized by dividing all elements of $h(r_k)$ by the total number of pixels in the image, denoted by n:

$$\begin{aligned} p(r_k) &= \frac{h(r_k)}{n} \\ &= \frac{n_k}{M \times N} \end{aligned} \qquad (2.8)$$

for $k = 0, 1, \cdots, L-1$. Note also that $n = M \times N$, where M and N are the row and column dimensions of the image. From basic probability, $p(r_k)$ gives the probability of occurrence of intensity level r_k in an image.

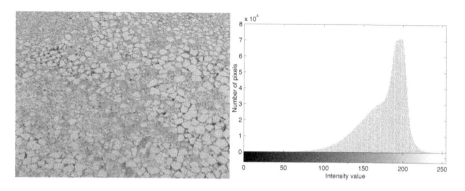

Figure 2.7 An 8-bit grayscale image and its histogram.

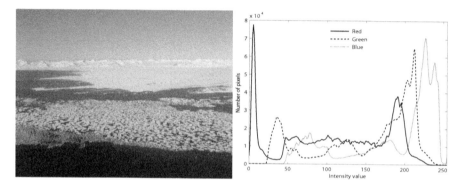

Figure 2.8 A color image and its histogram.

Histogram plays a basic role in image processing. It has been widely used in many areas such as enhancement, compression, segmentation, and description. In the image segmentation, the histogram can for instance be used to decide the value of threshold when converting a grayscale image into a binary image by the thresholding method (this will be introduced in Section 3.1).

2.3 BASIC RELATIONSHIPS BETWEEN PIXELS

2.3.1 PIXEL NEIGHBORHOODS

The neighborhood of a pixel plays an important role in image processing; it is often required for many operations, such as denoise, interpolation, edge detection, and morphology etc. The 4-neighbors and 8-neighbors are two common pixel neighborhoods that are used to process an image.

The 4-neighbors of a pixel p located at (x,y) are a set of pixels that connected vertically and horizontally to p. As seen in Figure 2.9(a), the 4-neighbors of p are denoted by $N_4(p)$, and given by:

$$(x+1,y),(x-1,y),(x,y+1),(x,y-1)$$

in terms of pixel coordinates. Each 4-neighbor of p is a unit distance from p.

The four pixels that connected diagonally to p are called diagonal neighbors (D-neighbors). As seen in Figure 2.9(b), the diagonal neighbors of p, denoted by $N_D(p)$, are given by:

$$(x+1,y+1),(x+1,y-1),(x-1,y+1),(x-1,y-1)$$

and each of them is at Euclidean distance of $\sqrt{2}$ from p.

The 8-neighbors of a pixel p include its four 4-neighbors and four diagonal neighbors as seen in Figure 2.9(c), and they are denoted by $N_8(p)$.

Be aware that some of the points in $N_4(p)$, $N_D(p)$, and $N_8(p)$ fall outside the image if p lies on the border of the image.

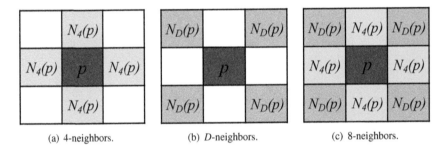

(a) 4-neighbors.   (b) D-neighbors.   (c) 8-neighbors.

Figure 2.9   Neighborhoods of a pixel.

2.3.2 ADJACENCY

Let V be a set of intensity values that is used to define adjacency. It specifies a criterion of similarity that the intensity values of adjacent pixels shall satisfy. For example, $V = 1$ when the adjacent pixels are 1-valued for a binary image. V could also be a subset of the 256 intensity values for an 8-bit grayscale image. Two pixels p and q with the intensity values from V are said to be:

(a) 4-adjacent, if $q \in N_4(p)$.
(b) 8-adjacent, if $q \in N_8(p)$.
(c) m-adjacent (mixed adjacent), if
    (i) $q \in N_4(p)$, or
    (ii) $q \in N_D(p)$ and $N_4(p) \cap N_4(q) = \emptyset$ (the set $N_4(p) \cap N_4(q)$ has no pixels whose intensity values are from V).

Mixed adjacency is a modification of 8-adjacency. It is used to eliminate the ambiguities that often arise when 8-adjacency is used (this will be explained in Section 2.3.3).

2.3.3   PATH

A path between pixels p_1 and p_n is a sequence of pixels $p_1, p_2, \cdots, p_n$ such that p_k is adjacent to p_{k+1}, for $1 \leq k < n$, and n is the length of the path. If $p_1 = p_n$, then the path is a closed path.

A path can be 4-, 8-, or m-paths depending on which form of adjacency is chosen. An example of 4-, 8-, or m-paths can be found in Figure 2.10. In this example, multiple paths (ambiguity) between two pixels are found in the case of 8-adjacency. To eliminate this ambiguity generated by the 8-adjacency, m-adjacency is used and only one path can be found between two pixels.

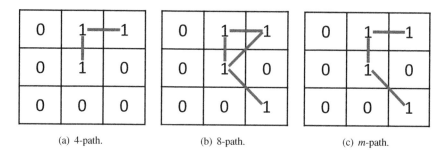

(a) 4-path.   (b) 8-path.   (c) m-path.

Figure 2.10   An example of 4-, 8-, or m-paths.

2.3.4   CONNECTIVITY

Two pixels p and q of an image subset S are said to be connected in S if there is a path from p to q consisting entirely of pixels in S. They are said to be 4-connected or 8-connected, depending on which path adjacency is used.

For any pixel p in S, the set of pixels that are connected to p in S is called a connected component of S. If S has only one connected component, then the set S is called a connected set [48]. The nature of a connected component depends on the type of adjacency that has been used. For example, a small binary image as seen in Figure 2.11(a) contains five connected components when using 4-adjacency. If using 8-adjacency, as shown in Figure 2.11(b), the number of connected components is reduced to two.

2.3.5   REGION AND BOUNDARY

A subset R of pixels in an image is called a region of the image if R is a connected set. Two regions R_1 and R_2 are said to be adjacent if their union forms a connected set, that is, if some pixel in R_1 is adjacent to some pixel in R_2. Regions that are not adjacent are said to be disjoint. When referring to regions, the type of adjacency should be specified (normally 4- and 8-adjacency are considered). Using 4-adjacency, then Figure 2.11(a) shows 5 disjoint regions, while the same image with 8-adjacency only has 2 disjoint regions as shown in Figure 2.11(b).

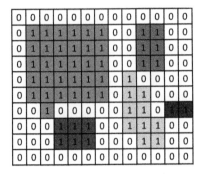

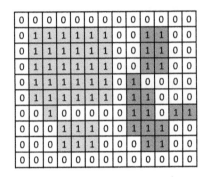

(a) Five 4-connected components.   (b) Two 8-connected components.

Figure 2.11   Connected components.

The boundary of a region R is the set of pixels that are adjacent to pixels in the complement of R. In other words, it is the set of pixels in the region for which one or more neighbors are not in R. The boundary of a finite region forms a closed path.

2.4   DISTANCE TRANSFORM

Distance transform is an important tool in image processing, and it is normally only applied to binary images that consist of object and background pixels. A distance transform of a binary image specifies the distance from every pixel to the nearest background pixel. In other words, the distance transform converts a binary image into a grayscale image where each object pixel has a value corresponding to the minimum distance from the background. The resulting grayscale image is a so-called distance map.

Assume f is a binary image, in which the pixels with a value of '0' indicate the background while the pixels with a value of '1' indicate the object. Let $B = \{p|f(p) = 0\}$ be the set of background pixels and $O = \{p|f(p) = 1\}$ be the set of object pixels. The distance transform of a binary image f, $D(p)$, can be given by [39]:

$$D(p) = \begin{cases} 0, & \text{if } p \in B \\ \min_{q \in B} d(p,q), & \text{if } p \in O \end{cases} \qquad (2.9)$$

where function d is a distance function or metric which is to determine the distance between pixels.

For pixels p, q, and r in an image, a distance function d satisfies the following three criteria [128]:

1. Positive definite: $d(p,q) \geq 0$ $(d(p,q) = 0$ iff $p = q)$
2. Symmetric: $d(p,q) = d(q,p)$
3. Triangular: $d(p,r) \leq d(p,q) + d(q,r)$

There are several types of distance metrics in image processing. The three most important ones are: Euclidean, city-block, and chessboard.

2.4.1 EUCLIDEAN DISTANCE

The Euclidean distance between two pixels p and q with coordinates (x,y) and (u,v), respectively, is defined as:

$$d_e(p,q) = \sqrt{(x-u)^2 + (y-v)^2} \tag{2.10}$$

Figure 2.12 shows the distance transform of a small binary image matrix by using the Euclidean distance metric. The pixels with the distance of some value r from the center pixel (x,y) form a circle centered at (x,y).

0	0	0	0	1	1	0
0	1	1	1	1	0	0
1	1	1	1	1	1	1
1	1	1	1	1	1	0
0	0	1	1	1	1	0
0	0	1	1	1	0	0
0	0	0	0	1	0	0

(a) Binary image matrix.

0.0000	0.0000	0.0000	0.0000	1.0000	1.0000	0.0000
0.0000	1.0000	1.0000	1.0000	1.0000	0.0000	0.0000
1.0000	1.4142	2.0000	2.0000	1.4142	1.0000	1.0000
1.0000	1.0000	1.4142	2.2361	2.0000	1.0000	0.0000
0.0000	0.0000	1.0000	2.0000	1.4142	1.0000	0.0000
0.0000	0.0000	1.0000	1.0000	1.0000	0.0000	0.0000
0.0000	0.0000	0.0000	0.0000	1.0000	0.0000	0.0000

(b) Distance transform of Figure 2.12(a) by using the Euclidean distance metric.

Figure 2.12   Distance transform.

2.4.2 CITY-BLOCK DISTANCE

The city-block distance (also known as the Manhattan distance) between two pixels p and q with coordinates (x,y) and (u,v), respectively, is defined as:

$$d_4(p,q) = |x-u| + |y-v| \tag{2.11}$$

where subscript "4" indicates that a 4-adjacency method is used when calculating the city-block distance. When using this distance metric, the pixels with the distance r from a center pixel (x,y) form a diamond centered at (x,y). For example, Figure 2.13(a) shows a city-block distance from a center pixel. The shape of distance value $r = 2$ is like a diamond, while the pixels with distance value $r = 1$ are the 4-neighbors of pixel (x,y).

2.4.3   CHESSBOARD DISTANCE

The chessboard distance between two pixels p and q with coordinates (x,y) and (u,v), respectively, is defined as:

$$d_8(p,q) = \max(|x-u|, |y-v|) \qquad (2.12)$$

where the subscript "8" indicates that an 8-adjacency method is used when calculating the chessboard distance. When using this distance metric, the pixels with the distance r from a center pixel (x,y) form a square centered at (x,y). For example, Figure 2.13(b) shows a chessboard distance from a center pixel. The shape of the distance $r=2$ is like a square, while the pixels with the distance $r=1$ are the 8-neighbors of the center pixel (x,y).

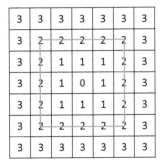

(a) City-block distance.   (b) Chessboard distance.

Figure 2.13   Distances from a center pixel.

2.4.4   PERFORMANCE OF THE DISTANCE METRICS

The Euclidean distance metric is an accurate distance that is often used. However, the Euclidean distance metric is time consuming due to its square, square-root, and floating-point computations. In contrast, the city-block and chessboard distance metrics can be computed much faster at the cost of being less accurate. Figure 2.14(a) gives a binary image of size 201×201 with 1-valued pixel at the center and 0-valued pixels everywhere else. The effects of different distance transforms on this binary image are shown in Figures 2.14(b) to 2.14(d).

2.5   CONVOLUTION

Convolution is an important operation that is widely used in image processing, such as smoothing, sharpening, and edge detection of images. In a 2-dimensional continuous space, the convolution of the impulse response $\omega(x,y)$ with a function $f(x,y)$,

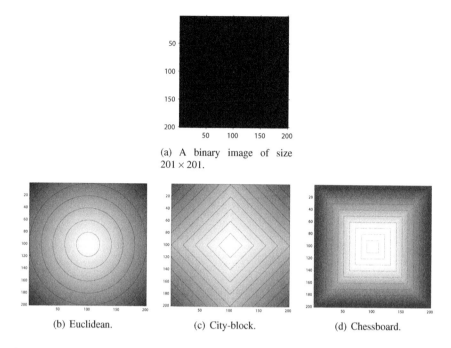

(a) A binary image of size 201×201.

(b) Euclidean.   (c) City-block.   (d) Chessboard.

Figure 2.14 Effects of different distance transforms.

denoted by the operator '$*$', is defined as:

$$h(x,y) = \omega(x,y) * f(x,y)$$
$$= \int_{-\infty}^{\infty} \int_{-\infty}^{\infty} \omega(u,v) f(x-u, y-v) du dv \tag{2.13}$$

In image processing, where an image is represented by a set of pixels, convolution is a local operation that replaces each pixel in an image by a linear combination of its neighbors. The impulse response $\omega(x,y)$ is then referred to as a convolution kernel, and the convolution becomes the calculation of the sum of products of the kernel coefficients with the intensity values in the region encompassed by the kernel. The convolution of a kernel $\omega(x,y)$ of size $m \times n$ with an image $f(x,y)$ is given by:

$$h(x,y) = \omega(x,y) * f(x,y)$$
$$= \sum_{s=-\frac{m}{2}}^{\frac{m}{2}} \sum_{t=-\frac{n}{2}}^{\frac{n}{2}} \omega(s,t) f(x-s, y-t) \tag{2.14}$$

For each pixel (x,y) in the image, the convolution value $h(x,y)$ is the weighted sum of the pixels in the neighborhood about (x,y), where the individual weights are the corresponding coefficients in the convolution kernel. This procedure involves translating the convolution kernel to pixel (x,y) in the image, multiplying each pixel in the neighborhood by a corresponding coefficient in the convolution kernel, and summing the multiplications to obtain the response at each pixel (x,y). Figure 2.15 gives an example of convolution of an image with a 3×3 kernel. In this example, the response of the kernel at the center point (x,y) of the 3×3 image neighborhood is given by:

$$
\begin{aligned}
h(x,y) =& \omega(-1,-1)f(x-1,y-1) + \omega(-1,0)f(x-1,y) \\
&+ \omega(-1,1)f(x-1,y+1) + \omega(0,-1)f(x,y-1) \\
&+ \omega(0,0)f(x,y) + \omega(0,1)f(x,y+1) + \omega(1,-1)f(x+1,y-1) \\
&+ \omega(1,0)f(x+1,y) + \omega(1,1)f(x+1,y+1)
\end{aligned}
\tag{2.15}
$$

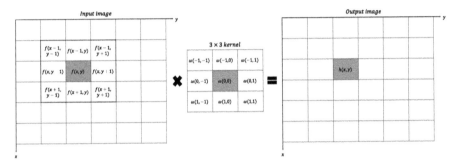

Figure 2.15 Convolution of an image with a 3×3 kernel.

The convolution kernels can be of any sizes. However, the kernels with odd sizes are most used in practice. This is because odd-size kernels have a unique center point and working with them is more intuitive.

2.6 SET AND LOGICAL OPERATIONS

Since a digital image is composed of pixels with particular values arranged in rows and columns, it can be represented as a set whose elements are the coordinates of the pixels, that is, ordered pairs of integers, and their intensity values. Then some basic concepts from set theory and logical operators are also applicable to digital images.

Set and logical operations have been widely used in image processing because of their simple and fast process. These operations are performed on a pixel-by-pixel basis between corresponding pixels of two or more images. That is, the value of a pixel in the output image depends only on the values of the corresponding pixels in the input images. Hence, the input images normally have to be of the same size.

2.6.1 SET OPERATIONS ON BINARY IMAGES

Since a binary image is a matrix containing object pixels of value 1 and background pixels of value 0, it can simply be represented as the set of those coordinate vectors

(x,y) of the pixels that have value of 1 in the binary image, given by:

$$G = \{(x,y) \mid g(x,y) = 1\} \tag{2.16}$$

where (x,y) are pairs of spatial coordinates, $g(x,y)$ is the pixel value (0 or 1) at (x,y), and G represents the set of image pixels describing the object of interest. All other image pixels are assigned to the background.

Let $\mathbb{Z}$ be the set of integers. Let the elements of a binary image be represented by a set $A \subseteq \mathbb{Z} \times \mathbb{Z}$, whose elements are 2-dimensional vectors of the form (x,y), which are spatial coordinates. If a set contains no elements, it is called an empty set or a null set, denoted by $\varnothing$. If $\omega = (x,y)$ is an element of A, then it is written as:

$$\omega \in A \tag{2.17}$$

otherwise, it is written as:

$$\omega \notin A \tag{2.18}$$

If every element of a set A is also an element of a set B, then A is said to be a subset of B and written as:

$$A \subseteq B \tag{2.19}$$

A set B of pixel coordinates ω that satisfy a particular condition is written as:

$$B = \{\omega \mid \text{condition}\} \tag{2.20}$$

The universe set, $\mathbb{U}$, is the set of all elements in a given application. In image processing, the universe is typically defined as the rectangle containing all the pixels in an image.

The complement (or inverse) of A, denoted as A^c, is the set of all elements of $\mathbb{U}$ that do not belong to set A, given by:

$$A^c = \{\omega \mid \omega \notin A\} = \mathbb{U} - A \tag{2.21}$$

The complement of the binary image A is the binary image that exchanges black and white, that is, 0-valued pixels set to 1-valued and 1-valued pixels set to 0-valued.

The union of two sets A and B, denoted as $A \cup B$, is the set of all elements that belong to either A, B, or both, given by:

$$A \cup B = \{\omega \in A \ \text{ or } \ \omega \in B\} \tag{2.22}$$

The union of two binary images A and B, is a binary image in which the pixels' values are 1 if the corresponding input pixels' values are 1 in A or in B.

Similarly, the intersection of two sets A and B, denoted as $A \cap B$, is the set of all elements that belong to both A and B, given by:

$$A \cap B = \{\omega \in A \ \text{ and } \ \omega \in B\} \tag{2.23}$$

The intersection of two binary images A and B is a binary image where the pixels' values are 1 if the corresponding input pixels' values are 1 in both A and B.

The difference of sets A and B, denoted as $A - B$, is the set of all elements that belong to A but not to B, given by:

$$A - B = \{\omega \in A \text{ and } \omega \notin B\} = A \cap B^c \tag{2.24}$$

which indicates the pixels in A that are not in B.

Figure 2.16 illustrates those basic set operations on binary images.

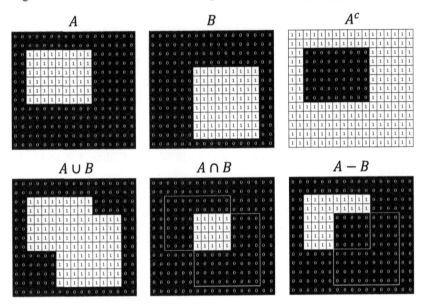

Figure 2.16 Basic set operations on binary images.

In addition to the preceding basic set operations, the concepts of set reflection and translation are used extensively in morphological image processing (which will be introduced in Section 4.2). The reflection of set A, denoted as $\hat{A}$, is to reflect all elements of A about the origin of this set, defined as:

$$\hat{A} = \{\omega \mid \omega = -a, \text{ for } a \in A\} \tag{2.25}$$

The reflection of an image A is the set of points in A whose (x, y) coordinates have been replaced by $(-x, -y)$. It is the reflected image of A across the origin (e.g., the version of A rotated $180°$ on the plane).

The translation of set A by point $z = (z_1, z_2)$, denoted as $(A)_z$, is to translate the origin of A to point z, defined as:

$$(A)_z = \{c \mid c = a + z, \text{ for } a \in A\} \tag{2.26}$$

The translation of an image A is the set of points in A whose (x, y) coordinates have been replaced by $(x + z_1, y + z_2)$. It is the translated image of A by placing the origin of A at z.

Figure 2.17 illustrates the operations of reflection and translation, in which the black dots identify the origins assigned to the sets.

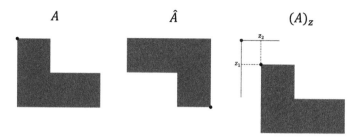

Figure 2.17 Reflection and translation of a set A.

2.6.2 SET OPERATIONS ON GRAYSCALE IMAGES

When dealing with grayscale images, the set must represent an image with pixels having more than two values. The image intensity value is the third dimension besides the two spatial dimensions x and y. A grayscale image can be represented as a binary image in a 3-dimensional space, with the third dimension representing image intensities. The intensity values can be viewed as heights at each pixel above the xy-plane, according to a function $z = g(x, y)$ corresponds to a surface in the 3-dimensional space. Thus, a grayscale image can be represented as a set given by:

$$G = \{(x, y, z) \mid z = g(x, y)\} \tag{2.27}$$

Because grayscale images are 3-dimensional sets, where the first two dimensions define the spatial coordinates and the third dimension denotes the grayscale intensity value, the preceding set operations for binary images are not applicable for grayscale images. Let the elements of a grayscale image be represented by a set $A \subseteq \mathbb{Z} \times \mathbb{Z} \times \mathbb{Z}$, whose elements are 3-dimensional vectors of the form (x, y, z), where the intensity value z is also an integer value within the interval $[0, 2^k - 1]$ with k defined as the number of bits used to represent z. The complement of A is defined as the pairwise differences between a constant and the intensity of every pixel in an image, given by:

$$A^c = \{(x, y, L - z) \mid (x, y, z) \in A\} \tag{2.28}$$

where $L = 2^k - 1$ is a constant. A^c is an image of the same size as A; however, its pixel intensities have been inverted by substracting them from the constant L.

The union of two grayscale sets (images) A and B is defined as the maximum of corresponding pixel pairs, given by:

$$A \cup B = \left\{ \max_z(a, b) \mid a \in A, b \in B \right\} \tag{2.29}$$

The outcome of $A \cup B$ is an image of the same size as these two images, formed from the maximum intensity between pairs of spatially corresponding elements.

Similarly, the intersection of two grayscale sets (images) A and B is defined as the minimum of corresponding pixel pairs, given by:

$$A \cap B = \left\{ \min_{z}(a,b) \mid a \in A, b \in B \right\} \qquad (2.30)$$

The outcome of $A \cap B$ is an image of the same size as these two images, formed from the minimum intensity between pairs of spatially corresponding elements.

2.6.3 LOGICAL OPERATIONS

The logical operations are derived from Boolean algebra, which is a mathematical approach to describe propositions whose outcome would be either TRUE or FALSE. The logical operations consist of three basic operations: NOT, OR, and AND. The terms NOT, OR, and AND are commonly used to denote complementation, union, and intersection, respectively. The NOT operation simply inverts the input value, that is, the output is FALSE if the input is TRUE, and it sets to TRUE if the input is FALSE. The OR operation produces the output TRUE if either one of the inputs is TRUE, and FALSE if and only if all the inputs are FALSE. The AND operation produces the output TRUE if and only if all inputs are TRUE, and FALSE otherwise. Any other logic operator, such as NAND, NOR, and XOR, etc., can be implemented by using only these three operators.

In image processing, the logic operations compare corresponding pixels of input images of the same size and generate an output image of the same size. When dealing with binary images, consisting of only 1-valued object pixels and 0-valued background pixels, the TRUE and FALSE states in logic operations correspond directly to the pixel values 1 and 0, respectively. Hence, the logic operations can be applied in a straight forward manner on binary images using the rules from logical truth tables, as shown in Table 2.1, to the pixel values from a pair of input images (or a single input image in the case of NOT operation).

Table 2.1

Truth tables of NOT, OR, and AND operations of all permutations of the inputs.

(a) NOT		(b) OR			(c) AND		
A	A^c	A	B	$A \cup B$	A	B	$A \cap B$
0	1	0	0	0	0	0	0
1	0	0	1	1	0	1	0
		1	0	1	1	0	0
		1	1	1	1	1	1

The logic operations can be extended to process the grayscale images whose pixel values are integers. In this case, logic operations are normally conducted in a bitwise arithmetic on the binary representation of pixels, comparing corresponding bits to produce the output pixel value. For instance, an 8-bit image has $2^8 = 256$ different grayscales. Let two input pixels have the values of 57 and 207 in decimal. Their respective binary representations are 00111001 and 11001111. Performing the AND operation on those two pixels in bitwise arithmetic results in the pixel with the binary value of 00001001 or decimal 9 at the corresponding position in the output image.

2.7   CHAIN CODE

Chain codes are a notation for recording the list of boundary pixels of an object. The chain code uses a logically connected sequence of straight-line segments with specified length and direction to represent the boundary [45]. A chain code can be created by tracking a boundary in some direction, say clockwise, and assigning a direction to the segments connecting every pair of pixels. The direction of each segment is coded by using a 4- or 8-connected numbering scheme, as shown in Figure 2.18. An example of the representations of an object boundary by using 4- and 8-directional chain codes are shown in Figure 2.19.

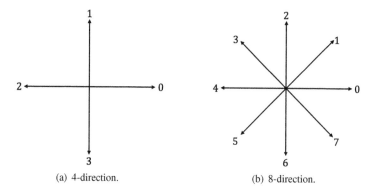

(a) 4-direction.   (b) 8-direction.

Figure 2.18   Numbering scheme of the chain code.

Taking an 8-connected numbering scheme, for example, each code indicates the change of angular direction (in multiples of $45°$) from one boundary pixel to the next. The even codes 0, 2, 4, and 6 correspond to horizontal and vertical directions, while the odd codes 1, 3, 5, and 7 correspond to the diagonal directions. The boundary has changed direction when a change occurs between two consecutive chain codes, and the change in the code direction usually indicates a corner on the boundary. By using the chain code, a complete description of an object boundary can be represented by the coordinates of the starting point together with the list of chain codes leading to subsequent boundary pixels, as shown in Figure 2.20. This representation of a list of boundary pixels becomes more succinct than using all boundary pixels' coordinates.

However, the chain code depends on the starting point, and different starting points result in different chain codes for the same boundary. To address this, the

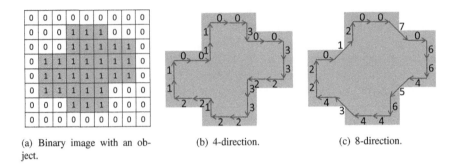

(a) Binary image with an object.

(b) 4-direction.

(c) 8-direction.

Figure 2.19 The chain codes for an object boundary.

chain code for a boundary can be normalized with respect to the starting point by treating it as a circular or periodic sequence of direction numbers, and redefining the starting point such that the resulting sequence of numbers is of minimum magnitude.

The chain code is not a rotation-invariant boundary description. A rotation of a boundary results in a different chain code. For example, if a boundary is rotated by $n \times 45°$, the new code for the rotated object is obtained by adding n mod 8 to the original code. A solution to this issue is to use the first difference of the chain code instead of the code itself. This difference reflects the spatial relationships between boundary segments and is independent of rotation. It is obtained by simply counting the number of direction changes (in counter-clockwise direction) that separate two adjacent elements of the code, and the last element of the difference is computed by using the transition between the first and last components of the chain. If the difference is less than 0, the difference should be modulo 4 or 8 in a 4- or 8-connected scheme, respectively. For example, let $C(i)$ be the chain code of the node i ($i = 0, 1, 2, \cdots, N-1$) on a boundary and N be the number of the nodes. The first difference chain code of the boundary in the 8-connected scheme, denoted as $D_1(i)$, can then be calculated by:

$$D_1(i) = (C(i+1) - C(i)) \mod 8, \quad i = 0, 1, 2, \cdots, N-2 \tag{2.31a}$$
$$D_1(N-1) = (C(0) - C(N-1)) \mod 8 \tag{2.31b}$$

Similar to the normalized chain code, the first difference can also be normalized by determining the sequence of numbers that forms an integer of minimum magnitude. Figure 2.21 illustrates the 8-directional chain codes, the first differences, and the normalized versions of the boundary in Figure 2.19.

It should be noted that chain code is not scale invariant; an object may have several chain codes at difference image resolutions. In addition, chain code is sensitive to noise along the boundary, as this may cause significant changes in the code and affect the principal shape features of the boundary.

Boundary

Representations

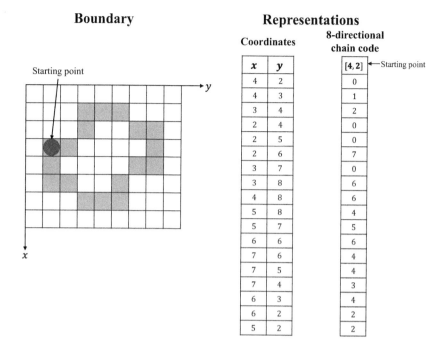

Figure 2.20 Representation for a boundary.

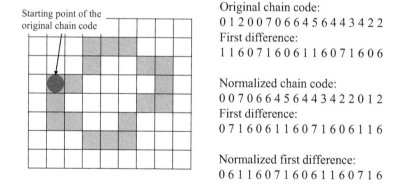

Original chain code:
0 1 2 0 0 7 0 6 6 4 5 6 4 4 3 4 2 2
First difference:
1 1 6 0 7 1 6 0 6 1 1 6 0 7 1 6 0 6

Normalized chain code:
0 0 7 0 6 6 4 5 6 4 4 3 4 2 2 0 1 2
First difference:
0 7 1 6 0 6 1 1 6 0 7 1 6 0 6 1 1 6

Normalized first difference:
0 6 1 1 6 0 7 1 6 0 6 1 1 6 0 7 1 6

Figure 2.21 A boundary's 8-directional chain codes and first differences.

2.8   IMAGE INTERPOLATION

An image gives the intensity values at the integral lattice locations, that is, the coordinates of each pixel are both integers. Image interpolation is the process of using known pixel intensity values to estimate the values at arbitrary locations other than those defined exactly by the integral lattice locations.

Image interpolation is a fundamental operation in image processing and has been widely used in image zooming, rotating, geometric calibration, etc. For example, as seen in Figure 2.22, suppose the input image coordinates (x, y) are assigned to another pair of image coordinates (η, ξ) by some coordinate transformation T:

$$(\eta, \xi) = T\{(x, y)\} \tag{2.32}$$

Then the intensity values of the input image also have to be assigned to the corresponding locations of the transformed image. However, with the coordinate transform T, some output pixels with coordinates calculated by Equation 2.32 may be located between the integer-valued grid points in the xy-plane. Thus, the image interpolation techniques are applied to determine the intensity values at those in-between locations. Note also that two or more pixels in the input image can be mapped into the same pixel in the output image by the coordinate transform, in which case the image interpolation techniques can be used to combine multiple input pixel values into a common output pixel value.

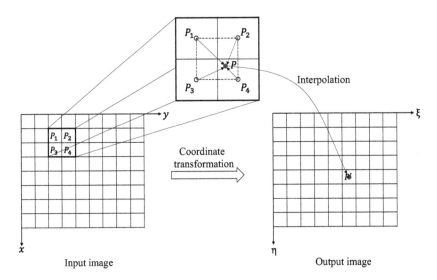

Figure 2.22   Image interpolation.

Many interpolation techniques have been developed. Among those, the nearest neighbor, bilinear, and bicubic interpolations are the three most popular methods.

2.8.1 NEAREST NEIGHBOR INTERPOLATION

The nearest neighbor interpolation, also called zero-order interpolation, assigns to each output pixel the intensity value of its nearest neighbor in the input image. To perform nearest neighbor interpolation method, the coordinates of every pixel in the output image, denoted as (m,n), are first mapped into the input image by:

$$(u,v) = T^{-1}\{(m,n)\} \qquad (2.33)$$

where (u,v) becomes the corresponding coordinates in the input image. Then the intensity value of the pixel located at (m,n) in the output image is set as the value of the pixel that has the shortest distance to (u,v) in the input image. This process is illustrated by Figure 2.23.

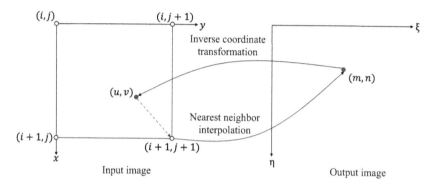

Figure 2.23 Nearest neighbor interpolation.

The nearest neighbor interpolation method is computationally very simple and fast. However, this method only uses the value of the pixel that is closest to the interpolated location, without taking account of the influence of other neighboring pixels. As a result, this method may produce severe mosaic and saw-tooth effect.

2.8.2 BILINEAR INTERPOLATION

The bilinear interpolation, also called first-order interpolation, calculates the intensity value for any point (u,v) in the input image by using a low-degree polynomial of the form:

$$f(u,v) = \sum_{m=0}^{1}\sum_{n=0}^{1} a_{mn} u^m v^n \qquad (2.34)$$

where the function f gives the intensity value at (u,v), a_{mn} $(m,n = 0,1)$ are coefficients determined by the four nearest neighbors.

When the intensity values of the four nearest neighbors are known, the general idea of the bilinear interpolation is to use linear interpolations along the x- and y-directions to determine the intensity value at (u,v). As exemplified in Figure 2.24, P denotes the interpolated point for which an intensity value must be calculated, (u,v)

are its coordinates mapped from the output image by Equation 2.33, and P_1, P_2, P_3, and P_4 are its four nearest neighbors in the input image with the coordinates (i, j), $(i, j+1)$, $(i+1, j)$, and $(i+1, j+1)$, respectively. The bilinear interpolation first interpolates linearly along the x-direction to find the values at Q_1 and Q_2:

$$f(Q_1) = (j+1-v)f(P_1) + (v-j)f(P_2) \qquad (2.35)$$

$$f(Q_2) = (j+1-v)f(P_3) + (v-j)f(P_4) \qquad (2.36)$$

then interpolates linearly along y-direction to obtain the value of P:

$$
\begin{aligned}
f(P) &= (i+1-u)f(Q_1) + (u-i)f(Q_2) \\
&= (i+1-u)\left[(j+1-v)f(P_1) + (v-j)f(P_2)\right] \\
&\quad + (u-i)\left[(j+1-v)f(P_3) + (v-j)f(P_4)\right] \\
&= (i+1-u)(j+1-v)f(P_1) + (i+1-u)(v-j)f(P_2) \\
&\quad + (u-i)(j+1-v)f(P_3) + (u-i)(v-j)f(P_4)
\end{aligned}
\qquad (2.37)
$$

which gives:

$$f(P) = \begin{bmatrix} i+1-u & u-i \end{bmatrix} \begin{bmatrix} f(P_1) & f(P_2) \\ f(P_3) & f(P_4) \end{bmatrix} \begin{bmatrix} j+1-v \\ v-j \end{bmatrix} \qquad (2.38)$$

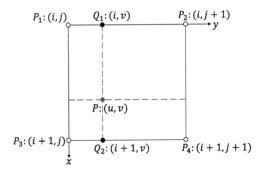

Figure 2.24 Bilinear interpolation.

In the bilinear interpolation, the intensity value of each output pixel is a weighted average value of its four nearest neighbors in the input image, and each weight is based on the distance from each of the four nearest pixels. As seen in Figure 2.24, the weights on each of the four nearest pixel intensities are actually proportional to the area of the opposing sub-rectangle. The closer the interpolated location is to an original coordinate in the input image, the larger the opposing sub-rectangle is, and the higher weight is assigned. This method is more complex than the nearest neighbor method, but it gives much better results than the nearest neighbor interpolation with a modest increase in computation time. It is typically a good choice for image interpolation.

2.8.3   BICUBIC INTERPOLATION

The bicubic interpolation, also called third-order interpolation, calculates the intensity value of any point (u, v) in the input image by reconstructing a surface among its four nearest neighbors based on their intensity values, the derivatives in both x- and y-directions, and the cross derivatives.

Similar to the bilinear interpolation, the bicubic interpolation calculates the intensity value for a point (u, v) by fitting a cubic polynomial:

$$f(u,v) = \sum_{m=0}^{3} \sum_{n=0}^{3} a_{mn} u^m v^n \tag{2.39}$$

where a_{mn} $(m, n = 0, 1, 2, 3)$ are coefficients determined by its 4×4 nearest neighbors in the input image, that is, the four nearest neighbors of the point (u, v) (empty circles as seen in Figure 2.25), and their horizontal, vertical, and diagonal neighboring pixels (black dots as seen in Figure 2.25). The latter are used to calculate the first-order derivatives in both x- and y-directions and the cross derivative at each of the four nearest neighbors of point (u, v). Then 8 first-order derivatives in both the x- and y-directions and 4 cross derivatives, together with 4 intensity values at the four nearest neighbors of point (u, v) give a linear system of 16 equations to determine the 16 coefficients of a_{mn} in Equation 2.39 [122].

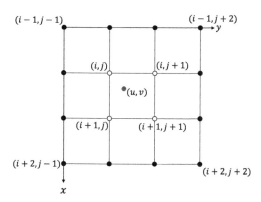

Figure 2.25   Bicubic interpolation.

Instead of directly calculating the solution of this linear system, typically by some matrix inversion, an alternative approach is to use a cubic convolution interpolation kernel that is composed of piecewise cubic polynomials defined on the subintervals $(-2, -1)$, $(-1, 0)$, $(0, 1)$, and $(1, 2)$ [78]. Assume the coordinates of the four nearest neighbors of point (u, v) in the input image are (i, j), $(i, j+1)$, $(i+1, j)$, and $(i+1, j+1)$. Then the interpolated pixel intensity may be expressed in the compact form [121]:

$$f(u,v) = \sum_{m=-1}^{2} \sum_{n=-1}^{2} f(u+m, v+n) r_c\{(m+i-u)\} r_c\{-(n+j-v)\} \tag{2.40}$$

where $r_c(x)$ denotes a bicubic interpolation function, given by :

$$r_c(x) = \begin{cases} (a+2)|x|^3 - (a+3)|x|^2 + 1, & \text{if } 0 \le |x| \le 1 \\ a|x|^3 - 5a|x|^2 + 8a|x| - 4a, & \text{if } 1 < |x| \le 2 \\ 0, & \text{if } |x| > 2 \end{cases} \qquad (2.41)$$

where a is the weighting factor that can be used as a tuning parameter to obtain a best visual interpolation result [118].

Compared with the bilinear interpolation, the bicubic interpolation method extends the influence of more neighboring pixels, and it takes not only the intensity values but also the intensity derivatives into account. Therefore, this method can produce more clear result than the bilinear interpolation method; however, at the expense of more computational complexity.

3 Ice Pixel Detection

Sea ice concentration is an important climatic and oceanic parameter that plays an important role in the studies of heat and moisture flux between ocean and atmosphere [137]. Besides, ice concentration has also been identified as one of the most influential factors on the magnitude of the loads on marine structures operating in the ice-covered Arctic (Antarctic) regions [164, 32]. As defined, ice concentration is given by a binary decision of each pixel to determine whether it belongs to the class "ice" or to the class "water". The distinction of the ice pixels from water pixels is thus crucial in order to obtain the ice concentration of an ice image.

In this chapter, two simple and popular image segmentation techniques, the Otsu thresholding method [114] and k-means clustering method [100] are introduced and compared to extract the ice pixels from ice images for the calculation of ice concentration. Both methods have been applied in the detection of the ice regions of various sea ice images [169, 155, 13, 40, 123, 73]. The applicable objectives and limitations of these two methods in ice concentration calculation are also discussed in this chapter.

3.1 THRESHOLDING

The pixels in the same region have similar intensity. Based on that ice is whiter than water, the pixel values are normally very different between ice and water pixels, and thresholding is thus a natural choice to segment ice regions from water regions.

The thresholding method is based on the pixel's grayscale value. It extracts the objects from the background and converts the grayscale image into a binary image. Assuming that an object is brighter than the background, the object and background pixels have intensity levels grouped into two dominant modes. The threshold T is selected to distinguish the objects from the background. A pixel is marked as "object" if its value is greater than the threshold value and as "background" otherwise, that is:

$$g(x,y) = \begin{cases} 1 & \text{if } f(x,y) > T, \\ 0 & \text{if } f(x,y) \leq T. \end{cases} \tag{3.1}$$

where $g(x,y)$ and $f(x,y)$ are the pixel intensity values located in the xth row, y^{th} column of the binary and grayscale image, respectively. This turns the grayscale image into a binary image.

When a constant threshold value is used over the entire image, it is called global thresholding. Otherwise, it is called variable thresholding, which allows the threshold to vary across the image.

3.1.1   GLOBAL THRESHOLDING

When the intensity distributions of objects and background pixels in an image are sufficiently distinct, it is possible to use a global threshold applicable for the entire image. The key to using the global thresholding is in how to select the threshold value, for which there are several different methods.

As mentioned in Section 2.2, image histogram is a useful tool for thresholding. If a histogram has a deep and sharp valley (local minima) between two peaks (local maxima), e.g., the bimodal histogram as shown in Figure 3.1, that represent objects and background, respectively, an appropriate value for the threshold will be in the valley between the two peaks in the histogram.

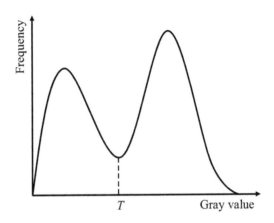

Figure 3.1   A bimodal histogram.

For example, as seen in Figure 3.2, the histogram of the grayscale sea ice image in Figure 3.2(a) clearly has two distinct modes, one for the objects (sea ice) and the other for the background (water). A suitable threshold for separating these two modes can be chosen at the bottom of this valley. As a result, it is easy to choose a threshold $T = 125$ that separates them. Then the grayscale image can be converted into the binary image as shown in Figure 3.2(c), and the ice concentration is thereby estimated as 41.47%.

This method is very simple. However, it is often difficult to detect the valley bottom precisely, especially when the image histogram is "noisy", causing many local minima and maxima. Often the objects and background modes in the histogram are not distinct, making it more difficult to determine where the background intensities end and the object intensities begin. Furthermore, in most applications there are usually enough variability between images such that, even if a global thresholding is feasible, an algorithm capable of automatically estimating the threshold value for each image will be most accurate.

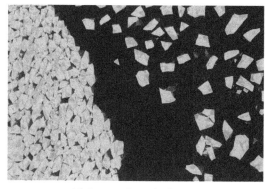

(a) A grayscale sea ice image.

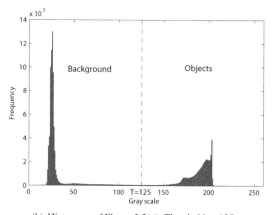

(b) Histogram of Figure 3.2(a). Threshold $= 125$.

(c) Binarized image of Figure 3.2(a), $IC = 41.47\%$.

Figure 3.2   Selecting a threshold by analyzing the peaks and valley of the histogram (Source of Figure 3.2(a): Figure from Q. Zhang, R. Skjetne, S. Løset and A. Marchenko, "Digital Image Processing for Sea Ice Observations in Support to Arctic DP Operations," In *ASME 31st International Conference on Ocean, Offshore and Arctic Engineering*, Rio de Janeiro, Brazil, 2012).

3.1.1.1   Otsu thresholding

To automatically select an optimal value for the threshold, Otsu proposed a method from the viewpoint of discriminant analysis; it directly approaches the feasibility of evaluating the "goodness" of the threshold [114].

Let $[0, 1, 2, \cdots, L-1]$ denote the L intensity levels for a given image with size $M \times N$, and let n_i denote the number of pixels with intensity i. The total number of pixels in the image, denoted by n, is then:

$$n = M \times N = \sum_{i=0}^{L-1} n_i \qquad (3.2)$$

To examine the formulation of this method, the histogram is normalized as a discrete probability density function:

$$p_i = \frac{n_i}{n}, \quad p_i \geq 0, \ \sum_{i=0}^{L-1} p_i = 1 \qquad (3.3)$$

Now suppose that a threshold t $(0 < t < L-1)$ is chosen to divide the pixels into two classes C_0 and C_1, where C_0 is the set of pixels with levels $[0, 1, \cdots, t]$, and C_1 is the set of pixels with levels $[t+1, t+2, \cdots, L-1]$. Then the probabilities of class C_0 occurrence is given by the cumulative sum:

$$P_0(t) = P(C_0) = \sum_{i=0}^{t} p_i \qquad (3.4)$$

Similarly, the probability of class C_1 occurrence is given by

$$P_1(t) = P(C_1) = \sum_{i=t+1}^{L-1} p_i = 1 - P_0(t) \qquad (3.5)$$

The mean intensity of the pixels in class C_0 is given by:

$$
\begin{aligned}
m_0(t) &= \sum_{i=0}^{t} i P(i|C_0) \\
&= \sum_{i=0}^{t} i \frac{P(C_0|i)P(i)}{P(C_0)} \\
&= \frac{1}{P_0(t)} \sum_{i=0}^{t} i p_i
\end{aligned}
\qquad (3.6)
$$

where $P(C_0|i) = 1$, $P(i) = p_i$, and $P(C_0) = P_0(t)$. Similarly, the mean intensity of the pixels in class C_1 is given by:

$$
\begin{aligned}
m_1(t) &= \sum_{i=t+1}^{L-1} i P(i|C_0) \\
&= \frac{1}{P_1(t)} \sum_{i=t+1}^{L-1} i p_i
\end{aligned}
\qquad (3.7)
$$

The average intensity (cumulative mean) up to the tth level is given by:

$$m(t) = \sum_{i=0}^{t} i p_i \tag{3.8}$$

and the average intensity of the entire image (global mean) is given by:

$$m_G = \sum_{i=0}^{L-1} i p_i \tag{3.9}$$

The validity of the following relation for any choice t can be easily verified:

$$P_0(t) m_0(t) + P_1(t) m_1(t) = m_G \tag{3.10}$$

and

$$P_0(t) + P_1(t) = 1 \tag{3.11}$$

The variances of pixel intensities in class C_0 and C_1 are given by:

$$\sigma_0^2(t) = \sum_{i=0}^{t} (i - m_0(t))^2 P(i|C_0)$$
$$= \frac{1}{P_0(t)} \sum_{i=0}^{t} (i - m_0(t))^2 p_i \tag{3.12}$$

$$\sigma_1^2(t) = \sum_{i=t+1}^{L-1} (i - m_1(t))^2 P(i|C_1)$$
$$= \frac{1}{P_1(t)} \sum_{i=t+1}^{L-1} (i - m_1(t))^2 p_i \tag{3.13}$$

respectively.

In order to evaluate the "goodness" of the threshold at level t, the following discriminant criterion measure is used in the discriminant analysis [114]:

$$\lambda = \frac{\sigma_B^2}{\sigma_W^2}, \quad \kappa = \frac{\sigma_G^2}{\sigma_W^2}, \quad \eta = \frac{\sigma_B^2}{\sigma_G^2} \tag{3.14}$$

where

$$\sigma_W^2(t) = P_0(t) \sigma_0^2(t) + P_1(t) \sigma_1^2(t) \tag{3.15}$$

is the within-class variance, that is, the weighted sum of the variances of each class,

$$\sigma_B^2(t) = P_0(t)(m_0(t) - m_G)^2 + P_1(t)(m_1(t) - m_G)^2$$
$$= P_0(t) P_1(t)(m_0(t) - m_1(t))^2$$
$$= \frac{(m_G P_0(t) - m(t))^2}{P_0(t)(1 - P_0(t))} \tag{3.16}$$

is the between-class variance, that is, a measure of separability between classes, and

$$\sigma_G^2 = \sum_{i=0}^{L-1} (i - m_G)^2 p_i \tag{3.17}$$

is the global variance, which is the intensity variance of all the pixels in the image. The relationship between the within-class variance, the between-class variance, and the global variance is:

$$\sigma_W^2(t) + \sigma_B^2(t) = \sigma_G^2 \tag{3.18}$$

The objective of the Otsu thresholding method is now to determine a threshold t that maximizes one of the criterion measures λ, κ or η.

Combing Equations 3.14 and 3.18, we obtain:

$$\lambda = \frac{\eta}{1 - \eta}, \qquad \kappa = \frac{1}{1 - \eta} \tag{3.19}$$

indicating that λ and κ are given in terms of η. The discriminant criteria for maximizing λ, κ and η with respective to t are then equivalent to one another.

Notice that σ_W^2 and σ_B^2 are functions of the threshold t, while σ_G^2 is a constant that is independent of t. Maximizing κ is equivalent to minimizing σ_W^2, maximizing η is equivalent to maximizing σ_B^2, and minimizing σ_W^2 is the same as maximizing σ_B^2. Therefore, the threshold with the minimum within-class variance also has the maximum between-class variance.

Since σ_W^2 is based on the second-order statistics (class variances) while σ_B^2 is based on the first-order statistics (class means), η is the simplest measure with respect to t. Thus, the Otsu thresholding method adopts η as the criterion measure to evaluate the "goodness" of the threshold t. By re-introducing t to η:

$$\eta(t) = \frac{\sigma_B^2(t)}{\sigma_G^2} \tag{3.20}$$

it follows that η also is a measure of class separability. The optimal threshold value, t^*, will then satisfy:

$$\sigma_B^2(t^*) = \max_{0 < t < L-1} \sigma_B^2(t) \tag{3.21}$$

Subject to the condition that $0 < P_0(t) < 1$, a maximum of $\sigma_B^2(t)$ always exists. A simple way to find t^* is to evaluate $\sigma_B^2(t)$ for all integer values of t, and find the maximum $\sigma_B^2(t)$. If there are more than one maximum, it is typical to average the corresponding values of t to obtain the final threshold.

Once t^* has been obtained, the maximum value $\eta(t^*)$ can be used as a quantitative measure to evaluate the separability of the two classes. This measure has value within the range:

$$0 \leq \eta(t^*) \leq 1 \tag{3.22}$$

The lower bound (zero) is attainable only when the image has a single constant intensity level, and the upper bound (unity) is attainable only when the image is 2-valued, that is the intensities are equal to 0 or $L - 1$, with $L = 2$.

Figure 3.3 shows the ice pixel detection result of Figure 3.2(a) by using the global Otsu thresholding method. In this case, the threshold is automatically chosen as $t = 108$, and the ice concentration is then estimated as 42.14%. The "goodness" of the threshold selected by Otsu's method, $\eta(t^*)$, is evaluated as $\eta(108) = 0.9643$, while it is evaluated as $\eta(125) = 0.9620$ when manually choosing 125 as the threshold.

Figure 3.3 Global threshold of Figure 3.2(a) by using Otsu thresholding method. Threshold $= 108$ and $IC = 42.14\%$ (Source: Figure from Q. Zhang, R. Skjetne, S. Løset and A. Marchenko, "Digital Image Processing for Sea Ice Observations in Support to Arctic DP Operations," In *ASME 31st International Conference on Ocean, Offshore and Arctic Engineering*, Rio de Janeiro, Brazil, 2012).

The Otsu thresholding method is an exhaustive algorithm of searching for the global optimal threshold. However, this method assumes that the histogram of the image is bimodal, and the performance of this method degrades rapidly as the object size decreases [82].

3.1.2 LOCAL THRESHOLDING

Global thresholding is simple and most effective in images with high levels of contrast. However, global thresholding may fail if some parts of the image are brighter (e.g., being illuminated by light) and some parts are darker (e.g., due to shadow). For example, when changing illumination across the scene of Figure 3.2(a) as seen in Figure 3.4(a), global thresholding fails to discriminate all ice from the water pixels, as shown in Figure 3.4(b).

Instead of using a single global threshold value, the local thresholding, which determines the thresholds locally, is typically required to handle uneven illumination problems. A common way is to divide the original image into sub-images and use different threshold values to segment each sub-image, as illustrated in Figure 3.4(c). In this manner, the overall ice concentration of the sea ice image, that in this case contains the factitious uneven illumination shown in Figure 3.4(a), is then estimated to 41.32%, which is close to the ice concentration derived from the original image by the global thresholding method.

Local thresholding has a better performance for images with uneven illumination, but this method induces difficulties, such as subdivision and subsequent threshold estimation.

3.1.3 MULTITHRESHOLDING

When several distinct objects are depicted within an image, more than one threshold is often required for proper segmentation. Multithresholding determines multiple threshold values for a given image and separates the image into several distinct regions, which correspond to one background and several objects. If k desired segments in an image can be distinguished by their gray values, where $T_1, T_2, \cdots, T_{k-1}$ $(0 < T_1 < T_2 < \cdots < T_{k-1} < L-1)$ are $k-1$ thresholds, the thresholded image can be given by:

$$g(x,y) = \begin{cases} g_k & \text{if } f(x,y) > T_{k-1}, \\ g_{k-1} & \text{if } T_{k-2} < f(x,y) \le T_{k-1}, \\ \quad \vdots & \\ g_2 & \text{if } T_1 < f(x,y) \le T_2, \\ g_1 & \text{if } f(x,y) \le T_1. \end{cases} \tag{3.23}$$

where $g_1, g_2, \cdots, g_k$ are the k valid intensity values, indicating k different segmented regions.

Multithresholding performs better than bi-level thresholding when the image has complex objects or background. The Otsu thresholding, which is a bi-level thresholding method, can be extended to multithresholding segmentation. In the case of k-thresholding with $k-1$ thresholds $t_1, t_2, \cdots, t_{k-1}$ $(0 < t_1 < t_2 < \cdots < t_{k-1} < L-1)$ for separating k classes: C_1 for $[0, 1, \cdots, t_1]$, C_2 for $[t_1 + 1, t_1 + 2, \cdots, t_2]$, $\cdots$, C_k for $[t_{k-1} + 1, t_{k-1} + 2, \cdots, L-1]$, the between-class variance is:

$$\sigma_B^2 = \sum_{j=1}^{k} P_j (m_j - m_G)^2 \tag{3.24}$$

where

$$P_j = \sum_{i \in C_j} p_i \tag{3.25}$$

$$m_j = \frac{1}{P_j} \sum_{i \in C_j} i p_i \tag{3.26}$$

and m_G is the global mean given in Equation 3.9. The optimal thresholds $t_1^*, t_2^*, \cdots, t_{k-1}^*$ can be computed as:

$$\sigma_B^2(t_1^*, t_2^*, \cdots, t_{k-1}^*) = \max_{0 < t_1 < t_2 < \cdots < t_{k-1} < L-1} \sigma_B^2(t_1, t_2, \cdots, t_{k-1}) \tag{3.27}$$

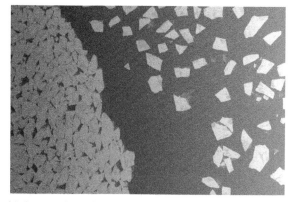

(a) A grayscale sea ice image containing factitious uneven illumination.

(b) Global Otsu thresholding result of Figure 3.4(a), threshold = 176.

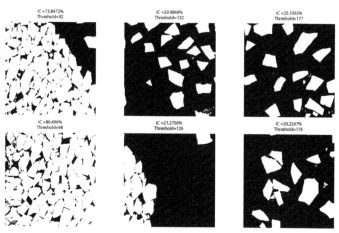

(c) Local Otsu thresholding result of Figure 3.4(a), $IC = 41.32\%$.

Figure 3.4 Global and local thresholding results of an uneven illumination sea ice image.

and the class separability degree can be measured as:

$$\eta(t_1^*, t_2^*, \cdots, t_{k-1}^*) = \frac{\sigma_B^2(t_1^*, t_2^*, \cdots, t_{k-1}^*)}{\sigma_G^2} \tag{3.28}$$

where σ_G^2 is the global variance given in Equation 3.17.

Figure 3.5 shows the 3-thresholding result of Figure 3.2(a) by Otsu's method. The two thresholds are determined as 61 and 142, and the coverages of each class can therefore be estimated as: 54.86% for black region (the region of open water in this case), 4.31% for gray region (the region of ice that has darker color in this case), and 40.83% for white region (the region of ice that has lighter color in this case).

Figure 3.5 3-thresholding result of Figure 3.2(a) by Otsu's method. Threshold 1 = 61, Threshold 2 = 142. The coverages of each segmented region: black region 54.86%, gray region 4.31%, and white region 40.83%.

Multithresholding using Otsu's method is very simple for $k = 2$ and 3. But as the number of classes increases, the maximization procedure becomes more computationally complex, which makes the Otsu multithresholding method time consuming.

3.2 CLUSTERING

Clustering is a technique for statistical data analysis. It is an unsupervised classification method that tries to find hidden structures in unlabeled data and assigns the unlabeled data into groups so that the data in one group are more similar to each other than to those in other groups.

The clustering algorithm is based on the measurement of mathematical distance between individual observations and groups of observations [44]. A distance measure is important in most clustering. This will determine the similarity of two calculated elements and affect the shape of the clusters. Distance in clustering can be in the Euclidean sense or non-Euclidean sense. A Euclidean distance is based on the locations of data points in Euclidean space, while a non-Euclidean distance is based on properties of data points rather than their locations in Euclidean space. Some common distance functions include the following:

1. Euclidean distance:

$$\|a - b\|_2 = \sqrt{\sum_i (a_i - b_i)^2} \tag{3.29}$$

2. Squared distance:

$$\|a - b\|_2^2 = \sum_i (a_i - b_i)^2 \tag{3.30}$$

3. Maximum norm distance:

$$\|a - b\|_\infty = \max_i |a_i - b_i| \tag{3.31}$$

4. City-block (Manhattan) distance:

$$\|a - b\|_1 = \sum_i |a_i - b_i| \tag{3.32}$$

5. Cosine distance (angle between two vectors):

$$\cos(a, b) = \frac{a^T b}{|a| \, |b|} \tag{3.33}$$

6. Mahalanobis distance:

$$d_M(a, b) = \sqrt{(a - b)^T S^{-1}(a - b)} \tag{3.34}$$

where S is the covariance matrix.

3.2.1 CLUSTERING TYPES

Based on whether or not the set of clusters is nested, the process of clustering can mainly be grouped as hierarchical or partition clustering [152]:

1. Hierarchical clustering finds successive clusters based on previously established clusters, and the set of nested clusters is organized as a hierarchical tree. Hierarchical clustering can be divided into two basic groups:
   a. Agglomerative clustering: bottom-up approach; start with each object as a separate cluster and then merge the objects into successively larger clusters.
   b. Divisive clustering: top-down approach; start with the whole set and proceed to divide it into successively smaller clusters.
2. Partition clustering determines all clusters at once, based on specifying an initialization of clusters, and iteratively reassigning objects among clusters until convergence is achieved. In partition clustering, the set of data objects is divided into non-overlapping clusters such that each data object is in exactly one cluster.

Clustering analysis is a good way for a quick review of data, especially if the objects are classified into many groups [8]. It has been widely used in many areas. Hierarchical clustering algorithms are popular in biological, social, and behavioral science because of the need to construct taxonomies, while partition clustering algorithms are used frequently in engineering applications where single partitions are important. Partition clustering is especially appropriate for the efficient representation and compression of large databases. Image segmentation, whose goal is to partition a given image into regions so that pixels belonging to one region are more similar to each other than pixels belonging to other regions, is closely related to the clustering problem. Because of the size of the pattern data (image matrix) and the need for a single partition, partition clustering algorithms are more popular for image segmentation than hierarchical algorithms [66]. Hence, we will introduce the k-means clustering, which is one of the most popular partition clustering algorithms, and its application in ice image segmentation.

3.2.2   K-MEANS CLUSTERING

The k-means clustering [100], where k is the number of desired clusters, is a simple and widely used partition clustering algorithm. It is a local optimal method that uses an iterative process to partition a set of data, $x_1, x_2, \cdots, x_n$, into k $(k < n)$ clusters, $S = \{S_1, S_2, \cdots, S_k\}$, by minimizing the within-cluster sum of squared distances (square-errors) to the cluster centroids. The k-means clustering algorithm uses the following objective function:

$$J = \sum_{i=1}^{k} \sum_{j=1}^{n_i} \left\| x_j^{(i)} - c_i \right\|^2 \tag{3.35}$$

where c_i $(i = 1, 2, \cdots, k)$ is the centroid or local mean of the data points $x_j^{(i)}$ in S_i, and n_i is the number of data points contained in S_i. The algorithm terminates when this objective function does not improve further.

The k-means clustering algorithm starts with k points, selected in some way (e.g., selected randomly from the data set, or specified by the user), as the initial cluster centroids. Then the algorithm iterates two steps: assignment and update [99]. In the assignment step, each point of the data set is assigned to its nearest centroid; and in the update step, the position of the centroid is adjusted to match the sample means of the data points that they are responsible for. The iteration stops when the positions of centroids no longer change.

Given a set of data, $x_1, x_2, \cdots, x_n$, the procedure of k means clustering can be expressed as follows:

Step 1: Select k points, $c_1, c_2, \cdots, c_k$ as the initial cluster centroids.
Step 2: Assign the data point x_p $(p = 1, 2, \cdots, n)$ to the cluster S_i $(i = 1, 2, \cdots, k)$ if it has shorter distance to the centroid c_i than any other centroid:

$$S_i = \left\{ x_p \mid \left\| x_p - c_i \right\|^2 \leq \left\| x_p - c_j \right\|^2, 1 \leq j \leq k, p \in [1, n] \right\} \tag{3.36}$$

Step 3: Update the positions of the k centroids by re-calculating the means of the data points assigned to that cluster:

$$c_i = \frac{1}{n_i} \sum_{j=1}^{n_i} x_j^{(i)} \qquad (3.37)$$

where n_i is the number of data points contained in S_i, and $x_j^{(i)}$ is the data point that belongs to S_i.

Step 4: Repeat Steps 2 and 3 until the positions of the centroids (local means) are unchanged.

Figure 3.6 gives an example to illustrate the process of the k-means clustering algorithm. In this example, to partition a given set of data points into two clusters, two initial centroid points are selected randomly from the data set. Each point in the data set is assigned to one of the clusters based on the minimum distance from the centroid. Then the centroids of each cluster are recalculated, and each data point is reassigned to the nearest centroid. Repeating the recalculations of the centroids and reassignments of the data points until the cluster membership does not change. Then the algorithm terminates.

In every iteration of the k-means clustering algorithm for a fixed number of clusters, the operations of reallocating each data point to the nearest centroid and setting cluster center as centroid minimize the sum of squared distances to centroids. This means that the objective function J decreases with each iteration of k-means clustering. Furthermore, the objective function J is lower-bounded by zero. Thus, the objective function J will stop decreasing at some point in a finite number of iterations, and then the k-means clustering algorithm converges. Usually, the k-means clustering algorithm has a rapid convergence. Figure 3.7 shows the application of the k-means clustering algorithm to ice pixel detection, where Figure 3.2(a) is divided into 2 and 3 clusters with the fact that 3 clusters give an IC that is a little bit higher than with 2 clusters.

From the procedure of the k-means clustering algorithm, it is clear to see that, although the k-means clustering is actually to minimize a within-class variance, it does not require to compute any variance. Therefore, this algorithm is computationally fast and tends to be used when large data sets are involved. However, the drawbacks of the k-means clustering algorithm are that it requires to initialize the proper number of clusters, and it is sensitive to the initial values for the centroid points.

How to choose the number k of clusters for any given data set is critical to the k-means clustering and has been one of the most difficult problems in data clustering. We usually do not know how many clusters exist for the data set, and there is no explicit mathematical criterion that can be evaluated to find it. Typically, the k-means clustering algorithm is performed independently for different values of k, and the partition that appears the most meaningful to the domain expert is selected [65].

Different initial centroid points can lead to different final clustering results. A good initialization of centroid points can lead to faster convergence and a better op-

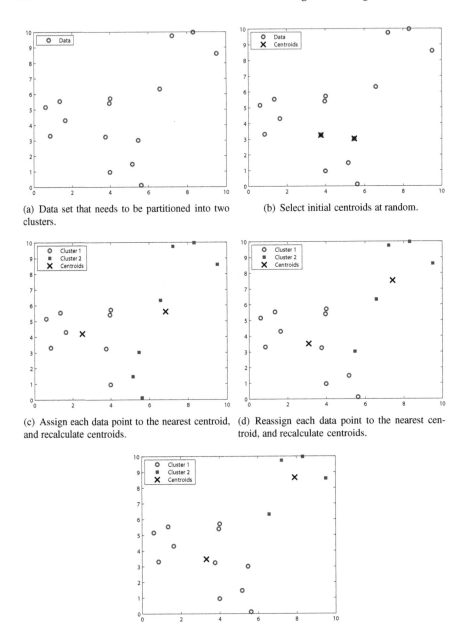

(a) Data set that needs to be partitioned into two clusters.

(b) Select initial centroids at random.

(c) Assign each data point to the nearest centroid, and recalculate centroids.

(d) Reassign each data point to the nearest centroid, and recalculate centroids.

(e) Repeat recalculating and reassigning until the clustering does not change.

Figure 3.6 An example of the *k*-means clustering algorithm process.

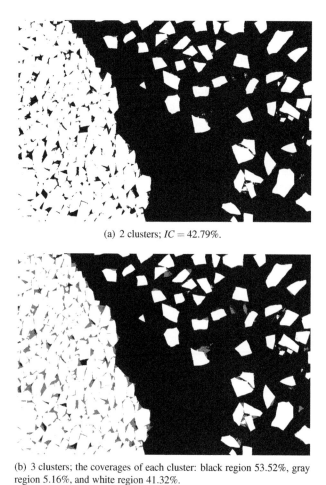

(a) 2 clusters; $IC = 42.79\%$.

(b) 3 clusters; the coverages of each cluster: black region 53.52%, gray region 5.16%, and white region 41.32%.

Figure 3.7 *K*-clustering results of Figure 3.2(a) (Source of Figure 3.7(b): Figure from Q. Zhang, R. Skjetne, S. Løset and A. Marchenko, "Digital Image Processing for Sea Ice Observations in Support to Arctic DP Operations," In *ASME 31st International Conference on Ocean, Offshore and Arctic Engineering*, Rio de Janeiro, Brazil, 2012).

timal solution, especially when the clusters are not separated well. This is because the k-means clustering algorithm, which is based on a square-error distance measure, can converge to local minima. There is thus no guarantee that this iterative algorithm will reach a global minimum. The most common way to overcome this issue is to perform the k-means clustering algorithm with a number of different initial centroid points. If they all lead to the same final clustering result, then we have some confidence that the global minimum of square-error has been achieved [66]. Otherwise, we can choose the partition with the smallest within-cluster sum of squared distances [65].

Furthermore, the k-means clustering algorithm is very sensitive to outliers, which are the data sufficiently "far away" from the rest of the data. An outlier that exists in the data set will be forced to belong to a cluster, and the shape of that cluster will be distorted as a result of this. As seen in Figure 3.8, for example, an outlier forces the k-means algorithm to put two compact and well-separated clusters into the same cluster, and the final cluster configuration is thereby changed. Thus, an outlier should be detected [62] and removed from the data set before running the k-means method.

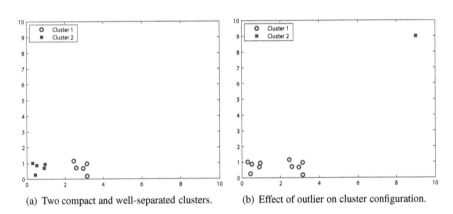

(a) Two compact and well-separated clusters. (b) Effect of outlier on cluster configuration.

Figure 3.8 An example showing an outlier affect final cluster configuration.

3.3 EXPERIMENT RESULTS AND DISCUSSION

In order to test the applicability of the (global) Otsu thresholding and the k-means clustering methods for sea ice observation, both methods are applied to the sea ice images obtained from the remote sensing mission at Ny-Ålesund in early May 2011 [185] to detect the ice pixels and calculate the ice concentration. In the k-means clustering, the equal division points are chosen as the initialization.

We first divide the image into only two clusters, where the first group represents sea ice, and the other group represents water. Parts of the processed images and ice concentration results are presented in Figures 3.9 to 3.11. The calculated ice concentrations are summarized in Table 3.1.

(a) Sea ice image 1.

(b) Otsu thresholding method, $IC = 15.36\%$.

(c) K-means clustering method with 2 clusters, $IC = 15.65\%$.

Figure 3.9 Sea ice image 1 and its ice pixel detection (Source: Figures from Q. Zhang, R. Skjetne, S. Løset and A. Marchenko, "Digital Image Processing for Sea Ice Observations in Support to Arctic DP Operations," In *ASME 31st International Conference on Ocean, Offshore and Arctic Engineering*, Rio de Janeiro, Brazil, 2012.)

(a) Sea ice image 2.

(b) Otsu thresholding method, $IC = 32.05\%$.

(c) K-means clustering method with 2 clusters, $IC = 32.49\%$.

Figure 3.10   Sea ice image 2 and its ice pixel detection (Source: Figures from Q. Zhang, R. Skjetne, S. Løset and A. Marchenko, "Digital Image Processing for Sea Ice Observations in Support to Arctic DP Operations," In *ASME 31st International Conference on Ocean, Offshore and Arctic Engineering*, Rio de Janeiro, Brazil, 2012).

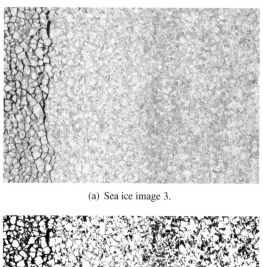

(a) Sea ice image 3.

(b) Otsu thresholding method, $IC = 72.63\%$.

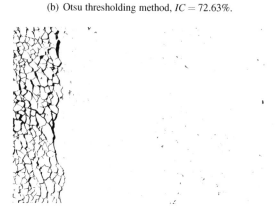

(c) K-means clustering method with 2 clusters, $IC = 96.50\%$.

Figure 3.11 Sea ice image 3 and its ice pixel detection (Source: Figures from Q. Zhang, R. Skjetne, S. Løset and A. Marchenko, "Digital Image Processing for Sea Ice Observations in Support to Arctic DP Operations," In *ASME 31st International Conference on Ocean, Offshore and Arctic Engineering*, Rio de Janeiro, Brazil, 2012).

Table 3.1
Ice concentrations of sea ice images.

Image no.	1	2	3
Otsu	15.36%	32.05%	72.63%
K-means	15.65%	32.49%	96.50%

When the intensity values of all the ice pixels in an image are significantly higher than water pixels, the ice pixel detection using the Otsu thresholding method is similar to the detection using the k-means method by dividing the image into two groups, as shown in Figures 3.9 and 3.10 and their calculated ice concentrations in Table 3.1. This is because the objective function of the k-means clustering method is equivalent to that of the Otsu thresholding method, and they are both based on an equivalent criterion that minimizes the within-class variance [87]. Compared with original sea ice images, the results obtained from applying both methods to calculate the ice concentration for Figures 3.9(a) and 3.10(a) are considered satisfactory.

However, as seen in Figure 3.11, the Otsu thresholding method can only find "light ice" pixels. The "dark ice" (e.g., brash ice, slush, and the ice that is submerged in water), whose pixel intensity values are closer to water pixels, may be lost. According to the definition of ice concentration given in Chapter 1, both "light ice" and "dark ice" visible in the ice images should be included when calculating the ice concentration. This will also be necessary for further identification of different types of sea ice (the details are described in Chapter 7).

In order to identify more ice pixels, the multi Otsu thresholding and k-means clustering methods can be applied to divide the image into three or more groups. Here we divided the sea ice image 3 into three groups, which we roughly interpret as "ice group 1" with the highest average pixel grayscale in white regions, water with the lowest average pixel grayscale in black regions, and "ice group 2" with the average pixel grayscale between in gray regions. The "ice group 1" region contains most of the ice floes in the image, and the "ice group 2" region contains most of brash ice and slush in the image. The coverage of each group is also calculated. The results are presented in Figure 3.12 and Table 3.2.

By comparing the calculated results for Figure 3.11(a) based on Otsu (Figure 3.11(b)), k-means clustering method with 2 clusters (Figure 3.11(c)), multi Otsu with 2 thresholds (Figure 3.12(a)), and k-means clustering method with 3 clusters (Figure 3.12(b)), it is the rather large content of brash ice and slush that results in the large difference between the ice concentration obtained from Otsu thresholding and k-means clustering methods. Since brash ice and slush are parts of the ice cover in the definition of ice concentration, we assume that an IC value of approximately $96 - 97\%$ is the correct range.

The multi Otsu thresholding and k-means clustering methods ensure a better de-

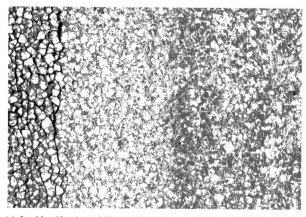

(a) Ice identification of Figure 3.11(a) by multi Otsu with 2 thresholds. $IC = 96.50\%$, "Ice group 1" in white: 54.39%, "Ice group 2" in gray: 42.11%, Water in black: 3.50%.

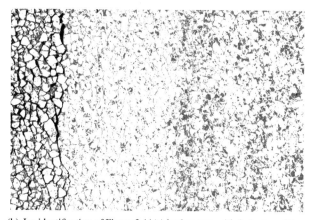

(b) Ice identification of Figure 3.11(a) by k-means with 3 clusters. $IC = 97.11\%$, "Ice group 1" in white: 77.91%, "Ice group 2" in gray: 19.20%, Water in black: 2.89%.

Figure 3.12 Ice pixel detection of Figure 3.11(a) by dividing the image into 3 groups (Source of Figure 3.12(b): Figure from Q. Zhang, R. Skjetne, S. Løset and A. Marchenko, "Digital Image Processing for Sea Ice Observations in Support to Arctic DP Operations," In *ASME 31st International Conference on Ocean, Offshore and Arctic Engineering*, Rio de Janeiro, Brazil, 2012).

Table 3.2

Image group coverages and ice concentrations for sea ice image 3 in Figure 3.11(a).

	Ice group 1	Ice group 2	Water	IC
Multi Otsu	54.39%	42.11%	3.50%	96.50%
K-means	77.91%	19.20%	2.89%	97.11%

tection by dividing the image into three or more groups, as shown in Figure 3.12(a) and 3.12(b). Additionally, the average intensity values for each segmented group of Figure 3.11(a) are calculated and shown in Table 3.3. It is found that the average intensity values for each group determined by the k-means clustering method are lower than the multi Otsu thresholding method. This implies that the k-means clustering method detects more ice pixels than the multi Otsu thresholding method, although both methods produce similar results in multilevel image segmentation [87]. Furthermore, the k-means clustering method is computationally faster than the Otsu thresholding method. Therefore, we conclude that the k-means clustering method is more effective than the (multi) Otsu method for the discrimination of ice pixels from water pixels when the image has a high ice concentration with a large amount of brash ice and slush.

Table 3.3

Average intensity values for each segmented image group.

	Ice group 1	Ice group 2	Water
Multi Otsu	218.1751	177.0690	63.8908
K-means	209,0405	161.6657	53.8234

However, to calculate ice concentration, both the Otsu thresholding and the k-means clustering methods divide an image into two or more classes in a mandatory manner. This will fail in the boundary conditions when ice concentration is 0% or 100%, which have to be dealt with as particular cases.

4 Ice Edge Detection

The size, shape, and location of the ice floes give important clues to their physical geometry and mechanics. The ice floe size distribution plays an important role in ice-structure analyses, and dynamic and thermodynamic processes. In image processing, the detection of individual ice floe boundaries is a key tool to accurately extract such information from ice images. A common approach for detecting object boundaries is to use edge detection, which can be further used in feature extraction, object location, and some properties such as area, perimeter, and shape measurements. This means that, with this technique, ice floe boundaries may be obtained to distinguish individual ice floes, and any geometric property of the ice floes together with the floe size distribution can thereby be estimated.

In this chapter, two common edge detection methods — derivative edge detection and morphology edge detection — are introduced. These methods are applied to sea ice images to try to extract ice floe boundaries and identify individual floes. The advantages and disadvantage of these two methods are also discussed.

4.1 DERIVATIVE EDGE DETECTION

Edges characterize object (or surface) boundaries that represent the change from one object (or surface) to another. This is useful for segmentation, registration, and identification of objects in a scene. In image processing, we identify the edges by detecting the difference between regions, which gives a rapid change in image brightness between neighboring pixels. For a rapid change, the first-order derivative (gradient) has a large magnitude and the second-order derivative (Laplacian) crosses zero, which are the two criteria that can be used to identify which pixels in an image may belong to an edge.

4.1.1 GRADIENT OPERATOR

The gradient, which is the first-order derivative, has a direction toward the most rapid change in intensity. The gradient of a digital image with pixel value $f(x,y)$ is defined as the vector:

$$\triangledown f = \begin{bmatrix} G_x \\ G_y \end{bmatrix} = \begin{bmatrix} \frac{\partial f}{\partial x} \\ \frac{\partial f}{\partial y} \end{bmatrix} \tag{4.1}$$

and the gradient magnitude is given by:

$$|\triangledown f| = \sqrt{G_x^2 + G_y^2} = \sqrt{\left(\frac{\partial f}{\partial x}\right)^2 + \left(\frac{\partial f}{\partial y}\right)^2} \tag{4.2}$$

59

while the direction of the gradient vector is given by the angle:

$$\theta = \angle f = \arctan\left(\frac{G_x}{G_y}\right) \tag{4.3}$$

with respect to the x-axis, where for implementation we use the arctan() function for correct quadrant mapping.

For computational efficiency, the gradient magnitude is sometimes approximated by using the squared gradient magnitude:

$$\nabla f \approx G_x^2 + G_y^2 \tag{4.4}$$

or the absolute gradient magnitude:

$$\nabla f \approx |G_x| + |G_y| \tag{4.5}$$

where these two approximations also preserve the relative changes in intensity scales.

The gradient of an image can be used for the detection of edges in the image; it requires the calculation of the partial derivatives G_x and G_y at every pixel location in the image. To directly estimate the partial derivatives G_x and G_y is one of the key issues in this method. The discrete approximation of partial derivatives over a neighborhood about a point is required. For example, it is a common and simple way to form the running difference of pixels along rows and columns of the image, which gives the approximation:

$$\frac{\partial f}{\partial x}(x,y) \approx f(x+1,y) - f(x,y) \tag{4.6a}$$

$$\frac{\partial f}{\partial y}(x,y) \approx f(x,y+1) - f(x,y) \tag{4.6b}$$

To implement the derivatives over an entire image, the edge detector, which is a local image processing method designed to detect edge pixels, filters the image with convolution kernels. So, the Equations 4.6a and 4.6b can then be implemented for all pertinent values of x and y by filtering $f(x,y)$ with the simple 1-dimensional convolution kernels shown in Figure 4.1.

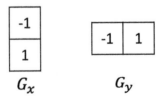

Figure 4.1   1-dimensional kernels used to implement Equations 4.6a and 4.6b.

The gradient approximations using 1-dimensional kernels are simple and efficient. However, using this method, G_x actually becomes the approximation to the gradient at the interpolated point $(x+\frac{1}{2},y)$ and G_y at $(x,y+\frac{1}{2})$. It is critical for the x and y partial derivatives to be computed at the exact position in space when computing an approximation to the gradient [67]. Therefore in practice, the kernels that are symmetric about the center point are used for gradient calculations, since this avoids the gradient to be calculated about an interpolated point between pixels. Furthermore, these kernels take into account the nature of the data on opposite sides of the center point, and thus they carry more information regarding the direction of an edge [48].

Many discrete differentiation operators are used for calculating an approximation of the gradient of the image intensity function, such as Sobel edge detector and Prewitt edge detector, which both use a 3×3 neighborhood (as shown in Figure 4.2(a)) for the gradient calculations. The Sobel edge detector calculates the gradient at the center point in a neighborhood by:

$$G_x(x,y) = [f(x+1,y-1) + 2f(x+1,y) + f(x+1,y+1)]$$
$$- [f(x-1,y-1) + 2f(x-1,y) + f(x-1,y+1)] \tag{4.7a}$$

$$G_y(x,y) = [f(x-1,y+1) + 2f(x,y+1) + f(x+1,y+1)]$$
$$- [f(x-1,y-1) + 2f(x,y-1) + f(x+1,y-1)] \tag{4.7b}$$

while the Prewitt edge detector calculates the gradient by:

$$G_x(x,y) = [f(x+1,y-1) + f(x+1,y) + f(x+1,y+1)]$$
$$- [f(x-1,y-1) + f(x-1,y) + f(x-1,y+1)] \tag{4.8a}$$

$$G_y(x,y) = [f(x-1,y+1) + f(x,y+1) + f(x+1,y+1)]$$
$$- [f(x-1,y-1) + f(x,y-1) + f(x+1,y-1)] \tag{4.8b}$$

Figures 4.2(b) and 4.2(c) show their edge detector kernels respectively. Note that the coefficients of all the kernels in Figure 4.2 sum to zero, thus giving a response of zero in areas of constant intensity, as expected of a derivative operator [48].

After the gradient is formed by the first-order difference, a common method to detect edges is by estimating the gradient magnitudes of the image at every point to generate a "gradient" image, and then thresholding this gradient image. Both the Sobel and Prewitt edge detectors combine their partial derivatives together to find the approximate absolute gradient magnitudes at each point in the grayscale image. Then this gradient magnitude of each pixel is compared to a threshold T to determine whether an edge exists. Correspondingly, the pixel at location (x,y) is considered as an edge pixel if its gradient magnitude exceeds the threshold T.

The threshold value determines the sensitivity of the edge detector. For noise-free images, the threshold can be chosen such that all amplitude discontinuities of a minimum contrast level are detected as edges. For noisy images, the threshold selection becomes a trade-off between missing valid edges and creating noise-induced false edges [121].

It should be noticed that the magnitude of the gradient is independent of the direction of the edge, so that both Sobel and Prewitt edge detectors are isotropic operators. The direction of the gradient (edge direction) is not as useful as the gradient

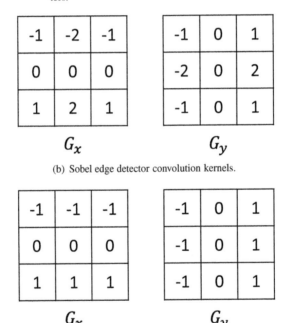

$f(x-1,y-1)$	$f(x-1,y)$	$f(x-1,y+1)$
$f(x,y-1)$	$f(x,y)$	$f(x,y+1)$
$f(x+1,y-1)$	$f(x+1,y)$	$f(x+1,y+1)$

(a) 3×3 image neighborhood for edge detectors.

-1	-2	-1
0	0	0
1	2	1

G_x

-1	0	1
-2	0	2
-1	0	1

G_y

(b) Sobel edge detector convolution kernels.

-1	-1	-1
0	0	0
1	1	1

G_x

-1	0	1
-1	0	1
-1	0	1

G_y

(c) Prewitt edge detector convolution kernels.

Figure 4.2   Examples of the edge detector kernels.

magnitude for the edge detection function, but it does complement the information extracted from an image using the magnitude of the gradient.

Although the Sobel and Prewitt edge detectors use slightly different kernels, they produce similar results. However, the Prewitt edge detector is slightly simpler to implement than the Sobel edge detector, whereas the Sobel edge detector is not as sensitive to noise as the Prewitt edge detector. This is because the coefficient with 2 in the Sobel operator provides extra smoothing, which makes the Sobel the preferable choice [48]. Figure 4.3 shows an example of derivative edge detection by using the Sobel and Prewitt edge detectors with the same thresholds. In this example, we see that the Sobel edge detection and the Prewitt edge detection give similar results.

(a) A grayscale sea ice floe image.

(b) Derivative edge detection of Figure 4.3(a) by using Sobel edge detector with a threshold $T = 0.05$.

(c) Derivative edge detection of Figure 4.3(a) by using Prewitt edge detector with a threshold $T = 0.05$.

Figure 4.3 Gradient edge detections Sobel and Prewitt operators.

4.1.2 LAPLACIAN

Similar to the first-order derivative, the second-order derivative, which is the Laplacian of the image, is defined as:

$$\nabla^2 f = \frac{\partial^2 f}{\partial x^2} + \frac{\partial^2 f}{\partial y^2} \tag{4.9}$$

The second-order derivative along the x direction can be approximated by differentiating Equation 4.6a with respect to x, e.g.:

$$\begin{aligned}
\frac{\partial^2 f}{\partial x^2}(x,y) &\approx \frac{\partial G_x(x,y)}{\partial x} \\
&= \frac{\partial f(x+1,y)}{\partial x} - \frac{\partial f(x,y)}{\partial x} \\
&\approx [f(x+2,y) - f(x+1,y)] - [f(x+1,y) - f(x,y)] \\
&= f(x+2,y) - 2f(x+1,y) + f(x,y)
\end{aligned} \tag{4.10}$$

Since this approximation is centered about the pixel $(x+1,y)$, however, we replace x with $x-1$ and obtain the result:

$$\frac{\partial^2 f}{\partial x^2}(x,y) \approx f(x+1,y) + f(x-1,y) - 2f(x,y) \tag{4.11}$$

This is the desired approximation to the second partial derivative centered about the pixel (x,y). Similarly,

$$\frac{\partial^2 f}{\partial y^2}(x,y) \approx f(x,y+1) + f(x,y-1) - 2f(x,y) \tag{4.12}$$

Combining Equations 4.11 and 4.12 two equations into a single operator according to Equation 4.9 gives an approximation of the Laplacian:

$$\nabla^2 f(x,y) = f(x-1,y) + f(x+1,y) + f(x,y-1) + f(x,y+1) - 4f(x,y) \tag{4.13}$$

This expression simply measures the weighted differences between a pixel and its 4-neighbors, and it can be implemented by using the kernel in Figure 4.4(a).

Sometimes it is desired to give more weight to the center pixels in the neighborhood, and Equation 4.13 can be extended to include the diagonal terms, for instance, using the kernel in Figure 4.4(b).

0	1	0
1	-4	1
0	1	0

(a) 4-neighbor kernel.

1	1	1
1	-8	1
1	1	1

(b) 8-neighbor kernel.

Figure 4.4 Examples of Laplacian convolution kernels.

The Laplacian edge detector is an isotropic operator, and it is sensitive to the changes in the gradient of the image. The response of the Laplacian can range from positive to negative. It is zero in the areas where the intensity of the image is constant or changes linearly, and it changes sign in the areas where the rate of image intensity change is greater than linear. The zero-crossing of the Laplacian indicates the presence of an edge.

However, zero-crossings do not always lie at edges. They can also occur in areas where the intensity of the image changes rapidly (e.g., the gradient of the image starts increasing or starts decreasing), and this may occur at places that are not associated with edges. A simple approach to the Laplacian zero-crossing detection is to form the maximum of all positive Laplacian responses and to form the minimum of all negative-value responses in a 3×3 window. If the magnitude of the difference between the maxima and the minima exceeds a threshold, then an edge is assumed to be present [121].

The Laplacian highlights regions of rapid intensity change. Therefore, the Laplacian edge detector is a useful tool for edge detection. However, because the Laplacian kernels approximate the second-order derivative measurement of the image, the Laplacian can be very sensitive to noise. Thus, the Laplacian is seldom used by itself for edge detection, but rather used in combination with other edge detection techniques. Moreover, in order to reduce sensitivity of the Laplacian filter to noise, an image is often smoothed first with a Gaussian filter before applying the Laplacian filter.

The Gaussian filter is a non-uniform lowpass filter. It is typically used to smooth (blur) an image. A 2-dimensional zero-mean Gaussian kernel is given by the function

$$G_\sigma(x,y) = \frac{1}{2\pi\sigma^2} e^{-\frac{x^2+y^2}{2\sigma^2}} \tag{4.14}$$

where σ is the standard deviation. The Gaussian kernel is rotationally symmetric; its coefficients decrease monotonically with increasing distance from the kernel's center. It theoretically requires an infinitely large convolution kernel for a Gaussian filter since the Gaussian distribution extends infinitely in the spatial domain. However, due to the fact that 3σ from the mean covers more than 99% of the Gaussian distribution, where the distribution is approximately zero more than 3σ distance, the discrete Gaussian kernel is typically approximated by removing the influence of points more than 3σ from the center pixel [111]. This means that σ determines the width of the Gaussian: a larger value of σ produces a wider Gaussian filter and greater smoothing (blurring) of the image. Figure 4.5(a) gives an example of a 5×5 convolution approximating the Gaussian function with $\sigma = 1$.

The combination of the Gaussian filter with the Laplacian filter, $\nabla^2 G_\sigma(x,y)$, is the so-called Laplacian of Gaussian (LoG). An expression for $\nabla^2 G_\sigma(x,y)$ is derived by the following differentiations:

$$
\begin{aligned}
\nabla^2 G_\sigma(x,y) &= \frac{\partial^2 G_\sigma(x,y)}{\partial x^2} + \frac{\partial^2 G_\sigma(x,y)}{\partial y^2} \\
&= \frac{1}{2\pi\sigma^2}\left[\frac{\partial}{\partial x}\left(-\frac{x}{\sigma^2}e^{-\frac{x^2+y^2}{2\sigma^2}}\right) + \frac{\partial}{\partial y}\left(-\frac{y}{\sigma^2}e^{-\frac{x^2+y^2}{2\sigma^2}}\right)\right] \\
&= \frac{1}{2\pi\sigma^2}\left[\left(\frac{x^2}{\sigma^4}-\frac{1}{\sigma^2}\right)e^{-\frac{x^2+y^2}{2\sigma^2}} + \left(\frac{y^2}{\sigma^4}-\frac{1}{\sigma^2}\right)e^{-\frac{x^2+y^2}{2\sigma^2}}\right] \\
&= \frac{1}{2\pi\sigma^2}\cdot\frac{x^2+y^2-2\sigma^2}{\sigma^4}e^{-\frac{x^2+y^2}{2\sigma^2}}
\end{aligned}
\tag{4.15}
$$

Thus, an LoG kernel can be approximated by sampling Equation 4.15 to the desired $n \times n$ size. However, a more effective approximation is to convolve an $n \times n$ Gaussian kernel with a Laplacian kernel (such as the kernel shown in Figure 4.4(a) or Figure 4.4(b)) [48]. Figure 4.5(b) gives an example of a 5×5 Laplacian of Gaussian kernel with $\sigma = 1$. Figure 4.6 shows an edge detection of Figure 4.3(a) by using a 13×13 LoG kernel with $\sigma = 2$ and a threshold $T = 0.005$.

0.0030	0.0133	0.0219	0.0133	0.0030
0.0133	0.0596	0.0983	0.0596	0.0133
0.0219	0.0983	0.1621	0.0983	0.0219
0.0133	0.0596	0.0983	0.0596	0.0133
0.0030	0.0133	0.0219	0.0133	0.0030

(a) A 5×5 Gaussian kernel.

0.0239	0.0460	0.0499	0.0460	0.0239
0.0460	0.0061	-0.0923	0.0061	0.0460
0.0499	-0.0923	-0.3182	-0.0923	0.0499
0.0460	0.0061	-0.0923	0.0061	0.0460
0.0239	0.0460	0.0499	0.0460	0.0239

(b) A 5×5 LoG kernel.

Figure 4.5 An example of 5×5 Gaussian and LoG kernels with $\sigma = 1$.

Figure 4.6 Laplacian of Gaussian edge detection of Figure 4.3(a), by using a 13×13 LoG kernel with a $\sigma = 2$ and a threshold $T = 0.005$.

4.2 MORPHOLOGICAL EDGE DETECTION

Morphology refers to geometrical characteristics related to the form and structure of objects, such as size, shape, and orientation. In image processing, mathematical morphology involves geometric analysis of shapes and textures in images based on some simple mathematical concepts from set theory. It is used to extract image components that are useful in representation and description of region shapes, such as boundaries, skeletons, convex hull, etc.

Morphological operators work with an image and a structuring element. The structuring element is a small set or subimage used to probe the given image for specific properties. It is also known as a kernel, and can be represented as a matrix of 0s and 1s. Values of 1 in the matrix indicate the points that belong to the structuring element, while values of 0 indicate otherwise. The structuring element has a desired shape, such as square, rectangle, disk, diamond, etc. The origin of a structuring element identifies the pixel of interest (the pixel being processed), and it must be clearly specified. The origin is typically at the center of gravity; however, it could be located at any desired position of the structuring element. Figure 4.7 shows examples of different structuring elements of various sizes with their origins highlighted in the corresponding geometric centers.

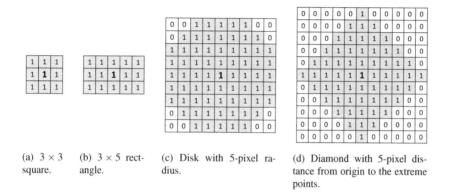

(a) 3 × 3 square. (b) 3 × 5 rectangle. (c) Disk with 5-pixel radius. (d) Diamond with 5-pixel distance from origin to the extreme points.

Figure 4.7   Examples of structuring elements with their highlighted origins.

Structuring elements consisting of a matrix of only 0s and 1s are called flat structuring elements. In a more complicated case, the points in the structuring element may have more values other than 0 and 1. Such structuring elements are called non-flat structuring elements. In practice, the non-flat structuring elements are rarely used, so they are not considered within the scope of this book, and all the structuring elements used in this book are flat, symmetrical, of unit height, with the origin at the geometric center.

When carrying out a morphological operation, the origin of the structuring element is translated to each pixel position in the image, and then the points within the translated structuring element are compared with the underlying image pixel values. If the translated structuring element matches a condition defined by the morphological operator in the image, the value of the pixel underneath the origin of the structuring element is modified. This happens through a comparison of the corresponding target pixel in the image with its neighbors. The comparison carried out in this process depends on which morphological operator is being used, and the effect of the morphological operation is determined by the structuring element used to process the image.

Before introducing the morphology-based edge detection method, some morphological operations are presented in this section. Those operations are the basic morphological operations that will be used in later chapters.

4.2.1   EROSION AND DILATION

Erosion and dilation are two fundamental operations in morphological image processing. Many other morphological operations, such as opening and closing, can be defined in terms of combinations of erosion and dilation in a sequence. By translating a structuring element to various points in the input image, the erosion and dilation operations examine the intersection (or union) between the translated kernel coordinates and the input image coordinates. The erosion operation "shrinks" or "thins" objects by removing pixels on object boundaries, while the dilation operation "grows" or "thickens" objects by adding pixels to the objects boundaries. The specific manner and amount of pixels added or removed are determined by the size and shape of the chosen structuring element.

4.2.1.1 Binary erosion and dilation

The erosion and dilation operations are basic components of a wide range of image processing algorithms, such as edge detection, noise removal, image segmentation, and image enhancement. These basic operations are particularly useful for the analysis of binary images.

Let $\mathbb{Z}$ denote the set of integers, A and B denote two sets in $\mathbb{Z} \times \mathbb{Z}$. The erosion of A by B, denoted by $A \ominus B$, is the set of all points z such that B translated by z is a subset of A, defined as:

$$A \ominus B = \{z \mid (B)_z \subseteq A\} \qquad (4.16)$$

where $(B)_z$ is the translation of B by the point $z = (z_1, z_2)$, defined as

$$(B)_z = \{b + z \mid b \in B\}, \qquad (4.17)$$

The erosion operation normally results in shrinking of the set A in a manner determined by the structuring element B.

If A is a binary image whose 1-valued pixels indicate object and 0-valued pixels indicate background, and B is a chosen structuring element, then the erosion of A by B is the set of all structuring element origin locations where the translated structuring element is entirely contained in the object of the binary image. To compute the erosion, the origin of the structuring element is translated throughout the domain of the binary image. For every translation of the structuring element, the pixel underneath the origin of the structuring element is set to 1 for the output image if the structuring element is entirely contained within an object subset (set of 1 values) of the input image; otherwise, it is set to 0. The process and the result of the erosion of a rectangular object (Figure 4.8(a)) by a cross-shaped structuring element (Figure 4.8(b)) are shown in Figures 4.8(c) and 4.8(d), respectively.

The dilation of A by B, denoted by $A \oplus B$, is the set of all points z such that the intersection of reflected and translated B with A is non-empty, defined as:

$$A \oplus B = \{z \mid (\hat{B})_z \cap A \neq \varnothing\} \qquad (4.18)$$

where $\hat{B}$ is the reflection of B, defined as

$$\hat{B} = \{w \mid -w \in B\}. \qquad (4.19)$$

The dilation operation normally results in an expansion of the set A in a manner determined by the structuring element B.

For a binary image A and structuring element B, the dilation of A by B is the set of all the reflected structuring element origin locations where the reflected and translated structuring element overlaps at least some portion of the object of the binary image. To compute the dilation, the reflected structuring element translates its origin throughout the domain of the binary image and checks to see if some part of it overlaps with object (1-valued) pixels in the binary image. If at least one point in the structuring element coincides with an object (1-valued) pixel in the binary image underneath, the pixel underneath the origin of the reflected structuring element in the output image is set to 1; otherwise, it is set to 0. The process and the result of the dilation of a rectangular object (Figure 4.8(a)) by a cross-shaped structuring element (Figure 4.8(b)) are shown in Figures 4.8(e) and 4.8(f), respectively. Note that the structuring element shown in Figure 4.8(b) is symmetrical with respect to its origin, that is $B = \hat{B}$.

Figure 4.9 gives the binary erosion and dilation results of Figure 4.3(a) by using a disk structuring element with 15-pixel-radius. It is obvious to see that the erosion shrinks the objects while the dilation thickens the objects.

4.2.1.2 Grayscale erosion and dilation

The basic operations of dilation and erosion can also be extended to grayscale images. Let f denote a grayscale image, and let b denote a flat structuring element. The erosion of f by b at any location (x,y) is defined as the minimum pixel value of the image subset that overlaps with b when the origin of b is at (x,y). The erosion at (x,y) of a grayscale image f by a structuring element b is then mathematically defined as:

$$[f \ominus b](x,y) = \min_{(s,t)\in b} \{f(x+s,y+t)\} \tag{4.20}$$

To compute the erosion of a grayscale image by a structuring element, the origin of the structuring element is placed at every pixel location in the image. Then the output pixel value at that location is the minimum value of the grayscale image over all values of the locations that overlap with the structuring element.

Conversely, the dilation of f by a structuring element b at any location (x,y) is defined as the maximum value of the image pixel values over the envelope outlined by the reflected $\hat{b} = b(-x,-y)$ when the origin of $\hat{b}$ is at (x,y), given by:

$$[f \oplus b](x,y) = \max_{(s,t)\in b} \{f(x-s,y-t)\} \tag{4.21}$$

To compute the dilation of a grayscale image by a structuring element, the origin of the reflected structuring element is placed at every pixel location in the image. Then the new dilated pixels value is set to the maximum value over the subset of the grayscale image formed by the region overlapped by the reflected structuring element.

Figure 4.10 gives the grayscale erosion and dilation results of Figure 4.3(a) by using a disk structuring element with 15-pixel-radius.

(a) A binary image with a rectangular object.

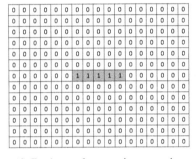

(b) A cross-shaped structuring element with the origin at center.

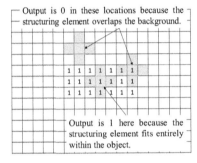

(c) Erosion process: structuring element translated to several locations on the image.

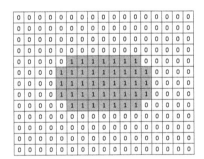

(d) Erosion result: output image matrix.

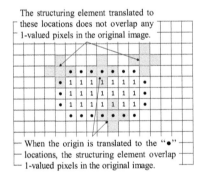

(e) Dilation process: structuring element translated to several locations on the image.

(f) Dilation result: output image matrix.

Figure 4.8   The processes of erosion and dilation [49].

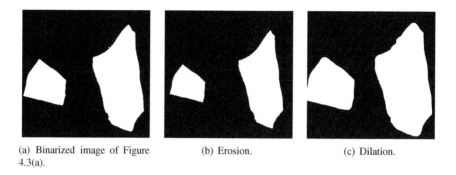

(a) Binarized image of Figure 4.3(a).　　　　(b) Erosion.　　　　(c) Dilation.

Figure 4.9 Binary erosion and dilation of Figure 4.3(a) by using a disk structuring element with 15-pixel-radius.

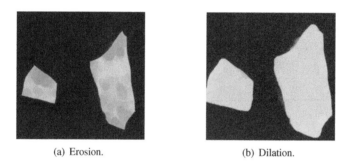

(a) Erosion.　　　　　　　(b) Dilation.

Figure 4.10 Grayscale erosion and dilation of Figure 4.3(a) by using a disk structuring element with 15-pixel-radius.

4.2.2 MORPHOLOGICAL CLOSING AND OPENING

4.2.2.1 Morphological closing

The morphological closing of A by B, denoted $A \bullet B$, is a dilation followed by an erosion, given by:

$$A \bullet B = (A \oplus B) \ominus B \qquad (4.22)$$

where A is an image, B is a structuring element, "$\oplus$" is the dilation operation, and "$\ominus$" is the erosion operation.

When A is a binary image, the closing is the complement of the union of all translations of B that do not overlap A [49]. All background regions in A that cannot contain the structuring element B are added to the set of object regions [146]. The binary closing operation tends to increase the spatial extent of the objects in an image. This is somewhat similar to the operation of dilation; however, the object shapes are better preserved. This is because the additional operation of erosion will mitigate the extension caused by the dilation when using the same structuring element.

When A is a grayscale image, understanding the grayscale closing operation requires treating $A(x, y)$ as a 3-dimensional surface where its intensity values are interpreted as heights over the xy-plane. Then the closing of A by B can be interpreted geometrically as pushing the structuring element B down on the top of the upper surface of A while translating B to all locations in A. At each location of the origin of B, the closing is the lowest value reached by any part of B as it slides over the upper surface of A. Figure 4.11 gives an example to illustrate the closing operation in one dimension, where the curve in Figure 4.11(a) is the intensity profile along a single row of an image. A structuring element is pushed down against the top of the curve in several locations, as seen in Figure 4.11(a). The solid curve in Figure 4.11(b) shows the result of the closing operation. As seen in Figure 4.11(b), the bottoms of valleys of the curve are clipped because the structuring element is too large to fit completely inside the downward valleys of the curve. The amount removed is proportional to how far the structuring element can reach into the valley. By using the closing operation, the dark details in A that are smaller than the structuring element B are suppressed, while the overall intensity levels and larger dark features are preserved [48, 49].

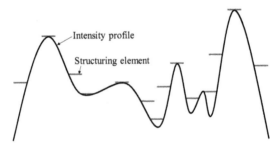

(a) 1-dimensional signal of intensity profile and structuring element pushed down along the top of the signal.

(b) The closing result of 1-dimensional signal in Figure 4.11(a).

Figure 4.11 An example illustrating the morphological closing in one dimension.

4.2.2.2 Morphological opening

The morphological opening of A by B, denoted $A \circ B$, is an erosion followed by a dilation, given by:

$$A \circ B = (A \ominus B) \oplus B \tag{4.23}$$

In binary morphology, where A is a binary image, the opening is the union of all the translations of the structuring element B that fit entirely within the binary image A [49]. All object regions in A that cannot contain the structuring element B are suppressed while the others are preserved [146]. The binary opening operation tends to decrease the spatial extent of the objects, somewhat like the erosion. However, it preserves the object shapes better than the erosion alone because the shrinking caused by the erosion will be mitigated by performing an additional dilation with the same structuring element.

In grayscale morphology where A is a grayscale image, one again treats $A(x,y)$ as a 3-dimensional surface. The opening of A by B can then be interpreted geometrically as pushing the structuring element B up from below the surface of A while translating B to all locations in A. The opening is constructed by finding the highest value reached by any part of B as it pushes up against the underside of A. Figure 4.12 illustrates this concept in one dimension. The tops of the peaks are clipped by the opening, where the amount removed is proportional to how far the structuring element is able to reach into the peak, as seen in Figure 4.12(b). By using the opening operation, the high intensity details in A that are smaller than the structuring element B are suppressed, while the overall intensity levels and larger bright features are preserved [48, 49].

4.2.3 MORPHOLOGICAL RECONSTRUCTION

Morphological reconstruction is often referred to as a geodesic [166]. It involves two images, called marker and mask, and a structuring element. The marker image is the starting point for the reconstruction. The mask image, which has the same definition domain as the marker image, restricts the reconstruction. And the structuring element defines the connectivity in the reconstruction [48].

4.2.3.1 Binary morphological reconstruction

Let G be the mask image, F be the marker image, and B be the structuring element. In the binary dilation-based morphological reconstruction, both mask and marker images are binary images, and the marker image F must be a subset of the mask image G, that is:

$$F \subseteq G \tag{4.24}$$

The geodesic dilation of size 1 of the marker image F by the structuring element B with respect to the mask image G, denoted by $D_G^{(1)}(F)$, is defined as:

$$D_G^{(1)}(F) = (F \oplus B) \cap G \tag{4.25}$$

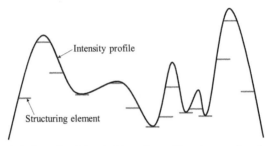

(a) 1-dimensional signal of intensity profile and structuring element pushed up along the undersurface of the signal.

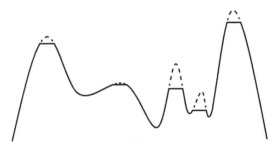

(b) The opening result of 1-dimensional signal in Figure 4.12(a).

Figure 4.12   An example illustrating the morphological opening in one dimension.

whereas the geodesic dilation of size n of F by B with respect to G is defined as:

$$D_G^{(n)}(F) = D_G^{(1)}[D_G^{(n-1)}(F)] \tag{4.26}$$

with $D_G^{(0)}(F) = F$.

Thus, the dilation-based binary morphological reconstruction is an iterative process by dilation with a marker image F by a structuring element B with respect to a mask image G until stability is obtained (the image no longer changes):

$$X_k = (X_{k-1} \oplus B) \cap G, \quad k = 1, 2, 3, \cdots \tag{4.27}$$

with $X_0 = F$, and terminating k when $X_k = X_{k-1}$. Then $R_G^D(F) = X_k = D_G^{(k)}(F)$ is the morphological reconstruction by dilation of G from F.

Similarly, the binary geodesic erosion of size 1 of F by B with respect to G is defined as:

$$E_G^{(1)}(F) = (F \ominus B) \cup G, \quad G \subseteq F \tag{4.28}$$

The geodesic erosion of size n of F by B with respect to G is defined as:

$$E_G^{(n)}(F) = E_G^{(1)}[E_G^{(n-1)}(F)] \tag{4.29}$$

with $E_G^{(0)}(F) = F$. Then the erosion-based morphological reconstruction is an iterative process of the geodesic erosion of F by B with respect to G until stability is achieved (the image no longer changes):

$$X_k = (X_{k-1} \ominus B) \cup G, \quad k = 1, 2, 3, \cdots \tag{4.30}$$

with $X_0 = F$, and terminating k when $X_k = X_{k-1}$. Then $R_G^E(F) = X_k = E_G^{(k)}(F)$ is the morphological reconstruction by erosion of G from F.

4.2.3.2 Grayscale morphological reconstruction

4.2.3.2.1 Geodesic dilation and dilation-based reconstruction

Let g be a grayscale mask image, f be a grayscale marker image, and b be a (flat) structuring element. In the grayscale geodesic dilation, the marker image f must be lower or equal to the mask image g:

$$f \leq g \tag{4.31}$$

that is, for each pixel (x, y) in the image domain, $f(x, y) \leq g(x, y)$.

The grayscale geodesic dilation of size 1 of the marker image by the structuring element with respect to the mask image, denoted by $D_g^{(1)}(f)$, is defined as:

$$D_g^{(1)}(f) = (f \oplus b) \wedge g \tag{4.32}$$

where $\wedge$ denotes the pointwise minimum operator, so that the geodesic dilation $D_g^{(1)}(f)$ is lower or equal to the mask image g: $D_g^{(1)}(f) \leq g$. The geodesic dilation of size 1 is obtained by first computing the dilation of the marker image f by the structuring element b, and then by choosing the minimum of the resulting dilated image and g at every pixel (x, y) [48]. The mask image g acts therefore as a limit to the propagation of the dilation of the marker image f [146].

The grayscale geodesic dilation of size n of f by b with respect to g is defined as:

$$D_g^{(n)}(f) = D_g^{(1)}[D_g^{(n-1)}(f)] \tag{4.33}$$

with $D_g^{(0)}(f) = f$.

The dilation-based grayscale morphological reconstruction is then also an iterative process of the geodesic dilation of a marker image f by a structuring element b with respect to a mask image g until stability is obtained (the image no longer changes):

$$X_k = (X_{k-1} \oplus b) \wedge g, \quad k = 1, 2, 3, \cdots \tag{4.34}$$

with $X_0 = f$, and terminating k when $X_k = X_{k-1}$. Then $R_g^D(f) = X_k = D_g^{(k)}(f)$ is the morphological reconstruction by dilation of g from f. The dilation-based grayscale morphological reconstruction extracts the peaks of the mask image that are marked by the marker image [166]. Figure 4.13 illustrates the dilation-based grayscale morphological reconstruction of a mask image from a marker image in one dimension.

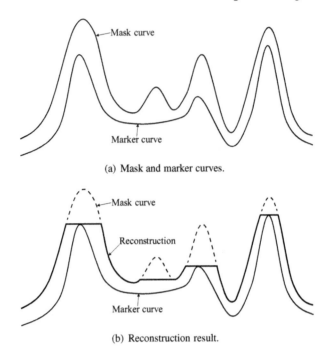

(a) Mask and marker curves.

(b) Reconstruction result.

Figure 4.13 An example of dilation-based grayscale morphological reconstruction in one dimension.

4.2.3.2.2 Geodesic erosion and erosion-based reconstruction

In the grayscale geodesic erosion, the marker image f must be greater or equal to the mask image g:

$$f \geq g \tag{4.35}$$

that is, for each pixel (x,y) in the image domain, $f(x,y) \geq g(x,y)$.

Similar to the grayscale geodesic dilation, the grayscale geodesic erosion of size 1 of f by b with respect to g is defined as:

$$E_g^{(1)}(f) = (f \ominus b) \vee g \tag{4.36}$$

where $\vee$ denotes the pointwise maximum operator, so that the geodesic erosion of an image remains greater or equal to its mask image. In the erosion case, the mask image acts as a limit to the shrinking of the marker image.

Then the geodesic erosion of size n of f by b with respect to g is defined as:

$$E_g^{(n)}(f) = E_g^{(1)}[E_g^{(n-1)}(f)] \tag{4.37}$$

with $E_g^{(0)}(f) = f$.

The erosion-based grayscale morphological reconstruction is an iterative process of the geodesic erosion of f by b with respect to g until stability is achieved (the image no longer varies):

$$X_k = (X_{k-1} \ominus b) \vee g, \quad k = 1, 2, 3, \cdots \tag{4.38}$$

with $X_0 = f$, and terminating k when $X_k = X_{k-1}$. Then $R_g^E(f) = X_k = E_g^{(k)}(f)$ is the morphological reconstruction by erosion of g from f. The erosion-based grayscale morphological reconstruction extracts the valleys of the mask image that are marked by the marker image. Figure 4.14 gives an example of the erosion-based grayscale morphological reconstruction of a mask image from a marker image in one dimension.

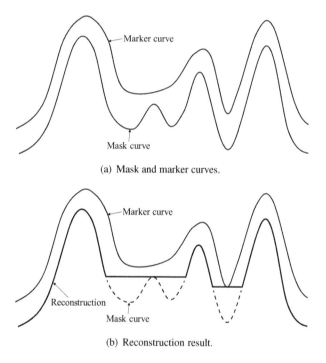

(a) Mask and marker curves.

(b) Reconstruction result.

Figure 4.14 An example of erosion-based grayscale morphological reconstruction in one dimension.

4.2.4 MORPHOLOGICAL GRADIENT

The erosion shrinks the objects in size and produces an interior of a point set, while the dilation expands the objects in size and produces an exterior of a point set. The erosion and dilation can be used in combination with image subtraction to emphasize the boundaries between regions.

Morphological gradients are based on the difference between extensive and anti-extensive transformations [126]. The dilation and erosion, with structuring elements containing their origin, correspond to extensive and anti-extensive operations, respectively. The basic morphological gradient, which is also called the Beucher gradient, is defined as the difference between the dilated and the eroded results of the image A with the structuring element B, given by:

$$\rho = (A \oplus B) - (A \ominus B) \tag{4.39}$$

For grayscale images, the basic morphological gradient results in an approximation of the absolute value of the gradient [12]. When the structuring element is relatively small, the homogeneous areas will not be affected much by dilation and erosion. In this case, the subtraction operation tends to eliminate these areas, and the result of the grayscale basic morphological gradient is an image with a "derivative-like (gradient)" effect [48].

The internal gradient is defined as the difference between the original image and the eroded image, given by:

$$\rho_{int} = A - (A \ominus B) \tag{4.40}$$

This gradient enhances the internal edges of the object. For binary images, the edges of the objects that are generated by the internal gradient are inside the objects.

Similarly, the external gradient is defined as the difference between the dilated image and the original image, given by:

$$\rho_{ext} = (A \oplus B) - A \tag{4.41}$$

This enhances the external edges of the object. For binary images, the edges of the objects, as generated by the external gradient, are outside the objects.

From Equations 4.39 to 4.41, we obtained:

$$\rho_{int} + \rho_{ext} = \rho \tag{4.42}$$

which shows that the basic morphological gradient enhances both internal and external edges, while the internal and external gradients are "half gradients" that are "thinner" than the basic morphological gradient. The internal and external gradients are used when thin contours are needed. The choice between internal or external gradient depends on the nature of the objects to be extracted [126].

Figures 4.15 and 4.16 show the binary and grayscale morphological edge detection results of Figure 4.3(a) respectively by using a disk structuring element with 15-pixel-radius. Those results confirm that the basic morphological gradients are wider than the internal and external gradients.

4.3 EXPERIMENTAL RESULTS AND DISCUSSION

The derivative and morphology edge detection methods have been applied to extract the boundaries of sea ice floes. As discussed in the previous sections:

a) The first derivative edge detection by using Sobel edge detector produces a similar edge detection result as Prewitt edge detector;

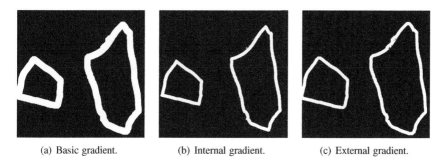

(a) Basic gradient.   (b) Internal gradient.   (c) External gradient.

Figure 4.15 Binary morphological edge detections of Figure 4.3(a) by using a disk structuring element with 157-pixel-radius.

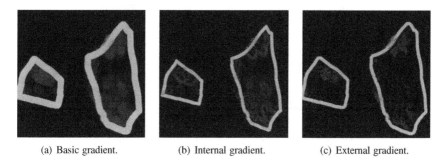

(a) Basic gradient.   (b) Internal gradient.   (c) External gradient.

Figure 4.16 Grayscale morphological edge detections of Figure 4.3(a) by using a disk structuring element with 15-pixel-radius.

b) The second derivative edge detection is very sensitive to noise;
c) Grayscale morphological gradient produces "derivative-like" edges.

Therefore, in this section, we only show the results using derivative edge detection by the Sobel detector and morphological edge detection by the binary internal gradient method. Their capacities for ice floe boundary detection is then discussed.

As seen from the sea ice floes on the right part in Figure 4.17, both the derivative and the morphology methods can extract the floe boundaries correctly when the ice floes are far away from each other. However, when the ice floes are close or connecting to each other, as seen for the floes in the left part of Figure 4.17, both methods struggle to find the edges between the floes. Thus, those methods are not sufficient, by themselves, to identify individual ice floes and calculating the size distribution.

Since the derivative method is used to identify rapid intensity changes in an image, more weak edge pixels between seemingly connected floes can be detected by the derivative method, as shown in Figure 4.18. However, the derivative method is more sensitive to noise than the morphology method. As shown in Figure 4.19, when decreasing the threshold for the derivative gradient, the method finds more edges,

(a) A sea ice image.

(b) Derivative edge detection by using Sobel edge detector with a threshold $= 0.05$.

(c) Morphology edge detection by using the internal gradient with a 5-pixel-radius disk structural element.

Figure 4.17   Edge detections of a sea ice image.

but at the cost of more noise. Moreover, the derivative method also produces more non-closed edges, especially in the detected weak edges between connected floes. Such non-closed edges indicate a loss of boundary information, so that separation of the connected floes becomes more difficult.

(a) A sample of connected ice floes from Figure 4.17(a).

(b) Edge detection of Figure 4.18(a) by the Sobel edge detector with the threshold of 0.05. Part of the weak edges (non-closed) between the connected floes are found.

(c) Edge detection of Figure 4.18(a) by the internal gradient morphology method with a disk structuring element of 5-pixel-radius. The detected edges are closed, but the weak edges between the connected floes are lost.

Figure 4.18   Comparison of the derivative and morphology methods.

(a) Sobel edge detection of Figure 4.18(a) with the threshold of $T = 0.05$.

(b) Sobel edge detection of Figure 4.18(a) with the threshold of $T = 0.03$. More weak edges (non-closed) between the connected floes are found at the cost of more noise.

Figure 4.19   Comparison of Sobel edge detection with different thresholds.

Since the morphology method is used to identify the difference between the extension and the anti-extension of a region's shape, a good description of the object's shape is given by this method. The morphology method is better in closing the edges than the derivative method. Moreover, since the erosion operation is to shrink objects, a thin connection between two floes can be broken when using the internal or basic morphology gradient with a proper structuring element, with the result that weakly connected ice floes can be separated.

Figure 4.20 shows the separation of the weakly connected sea ice floes in Figure 4.18(a) by the morphology method using the internal gradient with different sizes of structuring element. In Figure 4.20(a), the detected edges are thin, and the connected ice floes cannot be separated when using a 5-pixel-radius disk-structuring element. When enlarging the structuring element to a 16-pixel-radius disk, the detected floe edges become thick enough to break the connections between the floes and separate the connected floes, as seen in Figure 4.20(b). However, this is a special case and not a general property. Due to the various sizes and shapes of the ice floes, it is difficult to find a proper structuring element for an ice image. Furthermore, larger structuring elements result in thicker edges. The size of the ice floe will decrease when the detected floe edges become thicker. Therefore, this method only works for ice floe separation when the floes are weakly connected. The method is less suitable when trying to separate strongly connected ice floes.

(a) The internal gradient of Figure 4.18(a) with a 5-pixel-radius disk-structuring element. The detected floe edges are thin, and ice floes are connected.

(b) The internal gradient of Figure 4.18(a) with a 16-pixel-radius disk-structuring element. The detected floe edges become thicker, and the weakly connected ice floes are separated.

Figure 4.20 Analysis of weakly connected ice floes by using the internal gradient with different sizes of structuring element.

These experimental results show that individual ice floes cannot be distinguished well by either derivative or morphology methods when the floes are tightly connected. The inaccuracy of the ice floe edge detection will result in missing boundary information or non-closed boundaries, and these weaknesses will affect the statistical results. Therefore, in order to generate a precise ice floe size distribution, the development of an effective algorithm for ice floe boundary detection, especially handling the connected floe separation, is a main challenge that will be addressed in the following chapters of this book.

5 Watershed-Based Ice Floe Segmentation

The watershed transform [167] is a morphological-based algorithm that is used in image segmentation. The applications include biomedical images [50, 125, 134], various landscape images [53, 60, 72, 10], and also sea ice images [144, 180, 176, 177, 17, 13], etc. The watershed transform can be used to separate connected objects in an image. Unlike the morphological operations of dilation and erosion, which need to tune the sizes and shapes of the structuring elements according to the properties of the connected objects, the watershed-based segmentation first checks the regional minima in the image and then performs the transformation based on these. Thus, the watershed transform is able to separate connected objects regardless of their sizes. Due to this advantage, the watershed transform and its improvements have been widely used to solve connected objects division problems with acceptable results, such as grain [151, 163] and cell nuclei images [23, 5, 24, 28].

In the actual sea ice covered environment, ice floes typically touch each other and possibly overlap with snow on top. The junctions are therefore often difficult to detect. However, the apparent connection between ice floes should be identified, as it challenges the boundary detection algorithms and seriously affects ice floe size analysis. Thus, how to separate the connected floes effectively is a main challenge to obtain the information about individual ice floes and floe size distribution. A remedy to this problem is to use the watershed transform to separate the connected ice floes. For example, Burns et al. [17] performed the watershed transform on the binarized SAR sea ice image data by using the distance transform. In the watershed-based segmentation, however, the number of segmented regions depends on regional minima in the image. Usually, there is more than one regional minimum for each object, and this will induce over-segmentation. Hence, over-segmentation is a major problem to be dealt with. Refinements, such as minima-combination [46], mark-controlled [178, 165, 24], and H-dome transform [166, 24, 102, 90] were adopted to reduce the over-segmentation caused by multiple spurious regional minima. Unfortunately, due to varieties of sea ice floe shapes and sizes, it is hard to automatically locate the correct regional minima or markers for each ice floe. Blunt et al. [13] removed brash ice by using morphological opening and erosion operators and then adopted the watershed transform to separate the connected sea ice floes. Due to an ineluctable over-segmentation problem, they used a manual QA/QC step to rapidly identify over-segmented floes and join them.

In this chapter, the aim is to separate the connected ice floes, especially between those where the junction lines are almost invisible in the image. The distance transform-based watershed segmentation [9] is applied, and a combination of concave detection and neighboring-region merging is proposed to automatically reduce the over-segmentation caused by the watershed.

5.1   WATERSHED SEGMENTATION

The watershed transform is based on a topography interpretation of the grayscale image, where the pixel intensity values of the grayscale image are interpreted as heights over the xy-plane such that the image becomes a 3-dimensional topographic surface, as seen in Figure 5.1. If we imagine water falling on this surface, the water would intuitively flow toward the different minima that lie at the end of the paths of steepest descent of the topographic surface. It would then collect in the catchment basins, each of which is a set of points of the surface where the steepest descent paths eventually reach a single minimum. The points on this surface from where the water would equally likely flow toward more than one catchment basin, form ridge lines on the topographic surface. These are termed watershed lines, separating different catchment basins [48]. The concepts of the topographic representation of the image intensities, the minima, catchment basins, and watershed lines are illustrated in one dimension in Figure 5.2, where the curve with peaks and valleys exemplify the intensity profile along a single row of a grayscale image.

Note that a minimum in a grayscale image is a connected and iso-intensive area of possibly more than one pixel, where the intensity value is strictly lower than on the neighboring pixels [167]. To avoid confusion, a minimum is referred to as regional if it is a connected component of pixels with intensity value h, such that every pixel in the neighborhood of this region has a value strictly higher than h. In contrast to the regional minimum, a local minimum is a pixel in a grayscale image if and only if its intensity value is smaller or equal to that of any of its neighbors. According to these definitions, every pixel belonging to a regional minimum is a local minimum, but the converse is not true [166].

In terms of image segmentation, the objective of the watershed transform is to find all watershed lines. An intuitive way to numerically solve the watershed segmentation is the immersion approach [167]. In this method, imagine that a hole is punched in each minimum of the the topographic surface. The surface is then immerged into water with a constant vertical speed and is flooded from below by water rising into the different catchment basins through the holes. During the flooding, when the rising water in the different catchment basins is about to merge due to further flooding, a dam is built along the lines on the surface where the water would merge, to avoid the merging. This process is continued until the maximum level of flooding, that corresponds to the highest intensity value in the image is reached. At the end of the process, each minimum is surrounded by the dams delineating its associated catchment basin, and only tops of the dams are visible above the water surface. Then, the set of these dams define the watershed lines separating different objects or regions in the image. This flooding process can be implemented by using the morphological dilation [48], as introduced in the following.

Let $I(x,y)$ be the grayscale image, h_{min} and h_{max} be the lowest and highest values of $I(x,y)$ respectively. The topographic surface of $I(x,y)$ will be flooded from the elevations $h = h_{min} + 1$ to $h = h_{max} + 1$.

(a) A grayscale ice image.

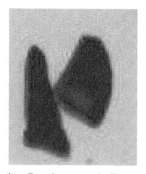

(b) Complement of Figure 5.1(a).

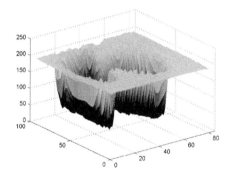

(c) Topographic surface of the grayscale image in Figure 5.1(b).

Figure 5.1 The topographic surface of a complemented grayscale ice image.

Denote $T_h(I)$ as the set of coordinates of the pixels in $I(x,y)$ lying below the elevation h, that is, the set of coordinates of the pixels standing for the threshold of $I(x,y)$ at level h, given by:

$$T_h(I) = \{(u,v)|I(u,v) < h\} \tag{5.1}$$

The minima is of primary importance in the watershed transform. According to the definition of the regional minimum given previously, the set of all regional minima of I at level h corresponds to the connected components of the cross-section of I at level h that are not connected to any component of the cross-section of I at level $h - 1$ [146]. Thus, a regional minimum M at level h of the grayscale image $I(x,y)$ is a connected component of $T_h(I)$ such that $M \cap T_{h-1}(I) = \varnothing$. In other words, a connected component M of $T_h(I)$ is a regional minimum at level h if and only if $M \cap T_{h-1}(I) = M \cap T_h(I + 1) = \varnothing$ [166]. Note that $T_{h-1}(I) = T_h(I + 1)$. Hence, an idea for extracting all regional minima of a grayscale image is given by using the

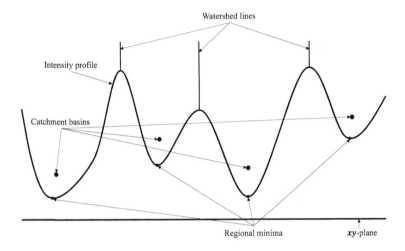

Figure 5.2 An example of illustrating the watershed segmentation in one dimension.

erosion-based grayscale morphological reconstruction:

$$M = R_I^E(I+1) - I \tag{5.2}$$

Let $M_1, M_2, \cdots, M_k$ be the sets of the coordinates of the pixels in the regional minimum of $I(x,y)$, let $C(M_i)$ denote the set of coordinates of the pixels in the catchment basin associated with the corresponding regional minimum M_i ($i = 1, 2, \cdots, k$), and let $C_h(M_i)$ denote the set of coordinates of pixels in the catchment basin associated with the regional minimum M_i at level h. The concepts of $T_h(I)$, $C(M_i)$, and $C_h(M_i)$ associated with the regional minimum M_i are explained in one dimension in Figure 5.3. As seen in this figure, the relationship among them can be found by:

$$C_h(M_i) = C(M_i) \cap T_h(I) \tag{5.3}$$

The intersection (AND) operator in this equation is simply used to isolate the portion of $T_h(I)$ that is associated with the regional minimum M_i at level h of flooding.

Let $C[h]$ be the union of the flooded catchment basins at level h, that is:

$$C[h] = \bigcup_{i=1}^{k} C_h(M_i) \tag{5.4}$$

Then $C[h_{\max} + 1]$ is the union of all catchment basins:

$$C[h_{\max} + 1] = \bigcup_{i=1}^{k} C(M_i) \tag{5.5}$$

The elements in both $C_h(M_i)$ and $T_h(I)$ are never replaced during execution of the algorithm. The number of elements in these two sets either increases or remains the

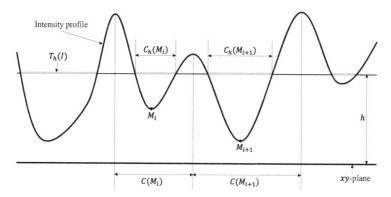

Figure 5.3 An example of the concepts of $T_h(I)$, $C(M_i)$, and $C_h(M_i)$ associated with the regional minimum M_i in one dimension.

same as h increase. Therefore, $C[h-1]$ is a subset of $C[h]$:

$$C[h-1] \subseteq C[h] \tag{5.6}$$

According to Equations 5.3 and 5.4, we then obtain that $C[h]$ is a subset of $T_h(I)$:

$$C[h] \subseteq T_h(I) \tag{5.7}$$

and, it follows that $C[h-1]$ is a subset of $T_h(I)$:

$$C[h-1] \subseteq T_h(I) \tag{5.8}$$

which indicates that each connected component of $C[h-1]$ is contained in exactly one connected component of $T_h(I)$. This relationship between $C[h-1]$ and $T_h(I)$ is important for obtaining $C[h]$ from $C[h-1]$ to find the watershed lines. Let S denote one of the connected components of $T_h(I)$. Then the construction of $C[h]$ from $C[h-1]$ depends on the following three possible relationships between each connected component $S \in C[h]$ and $C[h-1]$, as illustrated by Figure 5.4:

1. $S \cap C[h-1] = \varnothing$: In this case, S is obviously a new regional minimum of $I(x,y)$ at level h, as seen in Figure 5.4(a). The new regional minimum is then punched and its corresponding catchment basin will be progressively filled up with water.
2. $S \cap C[h-1] \neq \varnothing$ and is connected: In this case, $S \cap C[h-1]$ contains only one connected component of $C[h-1]$, and S belongs to the catchment basin associated with the regional minimum of $C[h-1]$, as seen in Figure 5.4(b).
3. $S \cap C[h-1] \neq \varnothing$ and is not connected: In this case, $S \cap C[h-1]$ contains more than one connected component of $C[h-1]$, and S contains different regional minima of $I(x,y)$ at level h, as seen in Figure 5.4(c). This condition occurs when part of a ridge separating two or more catchment basins appears at level h. The water level in these catchment basins would merge due to further flooding. Thus, a dam (or dams) must be built within S to prevent the merging between these

catchment basins. By dilating $S \cap C[h-1]$ with a 3×3 square structuring element (constraining the dilation to S), a 1-pixel-thick dam can be constructed.

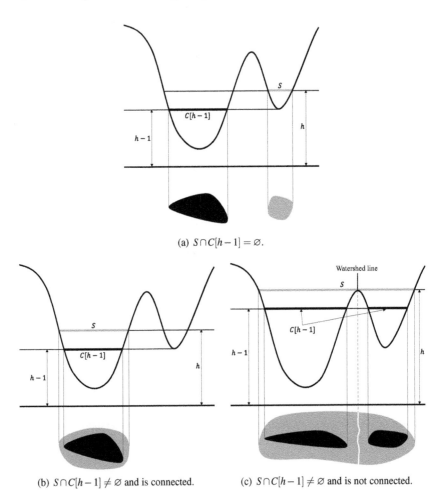

(a) $S \cap C[h-1] = \emptyset$.

(b) $S \cap C[h-1] \neq \emptyset$ and is connected. (c) $S \cap C[h-1] \neq \emptyset$ and is not connected.

Figure 5.4 The possible relationships between S and $C[h-1]$.

The algorithm for determining the watershed lines of grayscale image $I(x,y)$ is thus a recursive process of computing $C[h]$ from $C[h-1]$, with the initialization with $C[h_{\min}+1] = T_{h_{\min}+1}(I)$. In practice, only the values on h that correspond to the existing intensity values in $I(x,y)$ are used in this algorithm. And these values, as well as the values of $h_{\min}$ and $h_{\max}$, can be determined from the histogram of $I(x,y)$.

One should be aware of the connectivity used in the watershed computation. When using 4-neighbor connectivity in the watershed, the relationship between the segmented regions is 4-connected, while each determined watershed line forms an 8-connected path. In contrast, when using 8-neighbor connectivity, the segmented regions are 8-connected neighborhoods, and each determined watershed line forms a 4-connected path. It is important to note that the watershed lines produced by use of the 3×3 square structuring element in the watershed algorithm, as just described, form 1-pixel-thick connected paths. This property eliminates the problem of broken segmentation lines, which will be beneficial to the further analysis of the boundaries of segmented objects.

Normally, the grayscale of the original image is not appropriate for the direct application of the watershed segmentation algorithm. Instead of the original image itself, the watershed transform is applied to another image for which the catchment basins are the objects or regions we want to identify in the original image. Therefore, the key concept of the watershed segmentation is to convert the original image into this so-called segmentation function (image). A direct computation of the watersheds on the segmentation function should then yield the watershed ridge lines that segment the objects or regions of interest in the original image. In practice, image gradients, distance maps, and markers are commonly used to generate the segmentation functions in the watershed segmentation [49].

5.1.1 WATERSHED SEGMENTATION USING GRADIENTS

The gradient magnitude, which has high pixel values along object edges and low pixel values elsewhere, is natural to use in the watershed transform for grayscale image segmentation. This is because the main criterion of the segmentation is the homogeneity of the intensity values of the objects present in the image [9]. In this case, the objects correspond to the catchment basins and their contours correspond to the watershed lines of the gradient image [11].

In this approach, either a derivative gradient or a morphological gradient, as described in Chapter 4, can be used to pre-process a grayscale image prior to using the watershed transform. Based on the gradient magnitude, this would ideally produce watershed lines, which are the highest ridge lines separating the regional minima, along object edges. Then, those watershed lines are superimposed onto the original (or binarized) image to separate the connected objects. This process is illustrated in Figure 5.5.

Due to the problem of noise or local irregularities in the gradient image, shown in Figure 5.5(a), the watershed transform ends up producing far too many watershed lines, as seen in Figures 5.5(c) and 5.5(d), that do not correspond to the two objects of interest. A remedy to this problem is to first smooth the gradient image. Figure 5.6(a) shows the result of the smoothed gradient image by using a 7×7 square structuring element in a close-opening operation. As seen in Figures 5.6(b) and 5.6(c), the over-segmentation is effectively improved. However, there are still some extraneous watershed lines that make it difficult to determine which catchment basins are actually associated with the objects of interest.

The main problem here is that the edges between two connected ice floes in a grayscale image are sometimes too weak to be visible. In this case, the gradient becomes unsuitable to be used in the watershed segmentation, an example of which is seen in Figure 5.7. This issue will be addressed in the following section.

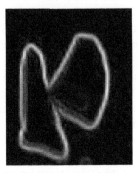

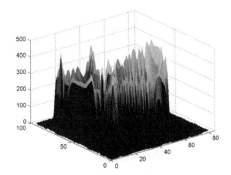

(a) Gradient image of Figure 5.1(a) by using 3 × 3 Sobel edge detection.

(b) Topographic surface of the gradient image in Figure 5.5(a).

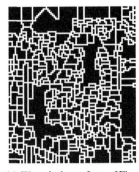

(c) Watershed transform of Figure 5.5(a).

(d) Over-segmented ice image by superimposing the watershed lines (in black) on to the original ice image.

Figure 5.5 Watershed segmentation based on gradient magnitude.

5.1.2 WATERSHED SEGMENTATION USING THE DISTANCE TRANSFORM

Often it is seen in an image that many objects overlap or connect to each other from their shape, but no edge is visible between them in the image. In this case, there is no contrast criterion that can be used to segment those seemingly overlapped or connected objects. To overcome this problem, the distance transform can be tested by building the so-called talweg lines. These are defined as the watershed lines of the inverted distance transform [11]. The approach is particularly helpful when the segmentation is based on the shape of the objects.

To use the distance transform, the watershed first treats the inverse distance map as a topographic surface, as shown in Figures 5.8(b) and 5.8(c). Then it checks the regional minima in the inverse distance map, which are actually the regional maxima

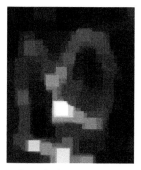 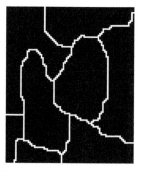 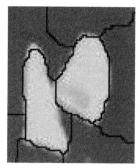

(a) Smoothed gradient image of Figure 5.5(a) by using a close-opening.

(b) Watershed transform of the smoothed gradient in Figure 5.6(a).

(c) Segmented ice image by superimposing the watershed lines (in black) on to the original ice image.

Figure 5.6 Watershed segmentation based on smoothed gradient magnitude.

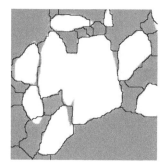

Figure 5.7 The effect of the gradient-based watershed segmentation to a grayscale ice image in which the floe edges are invisible.

of the distance map, as shown in Figure 5.8(d). Based on these regional minima, the watershed performs its transformation to find the watershed lines, as seen in Figure 5.8(e). Superimposing those watershed lines on the binary image results in the segmented image, as shown in Figure 5.8(f).

The number of segmented regions depends on the regional minima found in the inverse distance map. Usually, there is more than one unique regional minimum for each object, and this will usually induce over-segmentation. In Figure 5.9, different distance transforms are tested for the watershed-based segmentation of the ice floe as in Figure 5.8(a). We see that an ice floe is over-segmented because multiple spurious regional minima are found by the tested Euclidean and the city-block distance transforms.

The Euclidean distance is most likely to cause over-segmentation, especially in the area between two connected objects. This is because the Euclidean distance propagates outwards in the shape of a circle. This makes it easy to form a small

(a) A binary ice image.

(b) Inverse chessboard distance map of Figure 5.8(a).

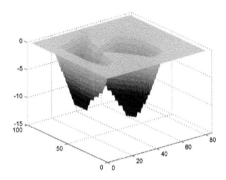

(c) Topographic surface of the chessboard distance map in Figure 5.8(b).

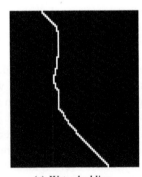

(d) Regional minima superimposed in black on the binary ice image.

(e) Watershed line.

(f) Segmented ice image by superimposing the watershed lines (in black) on the binary image.

Figure 5.8   Watershed segmentation based on distance transform.

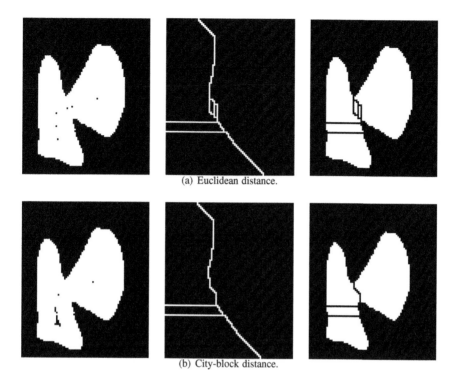

(a) Euclidean distance.

(b) City-block distance.

Figure 5.9 Watershed segmentation using different distance transforms. Left: Regional minima superimposed in black on the binary ice image. Middle: Watershed lines. Right: Resulting segmented ice image.

"island" made of a few pixels between different objects, as seen in the middle of Figure 5.10(b). This small "island" will be treated as a separate minimum when performing the watershed transform. The city-block distance, on the other hand, has a moderate probability of over-segmentation due to its diamond shape propagation. This has a tendency to form multiple regional minima pixels with different gray scales in the area near the center of the object, where those also may lead to over-segmentation. Finally, the chessboard distance has the lowest probability of over-segmentation compared to the Euclidean and city-block distances, due to its square shape propagation. However, the chessboard distance might be unable to separate the two connected objects if a high percentage of their boundary are connected [22]. As a matter of fact, the chessboard distance tends to cause under-segmentation, which is another critical issue in the watershed segmentation. Figure 5.11 gives an example that shows the effects of the watershed segmentations based on different distance transforms.

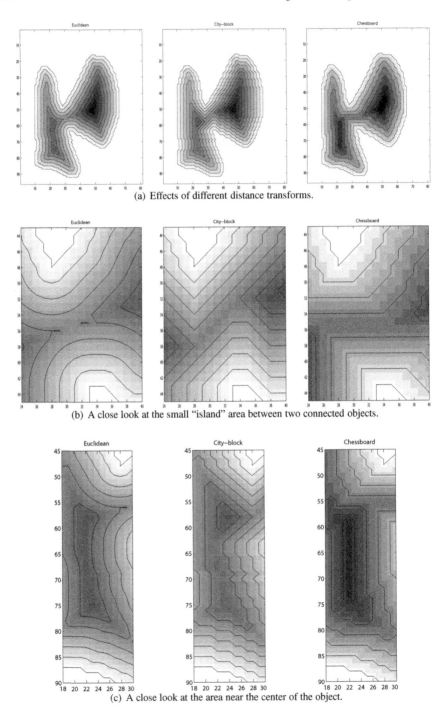

(a) Effects of different distance transforms.

(b) A close look at the small "island" area between two connected objects.

(c) A close look at the area near the center of the object.

Figure 5.10   Comparison of different distance transforms on the binary ice image shown in Figure 5.8(a). Left: Euclidean. Middle: City-block. Right: Chessboard.

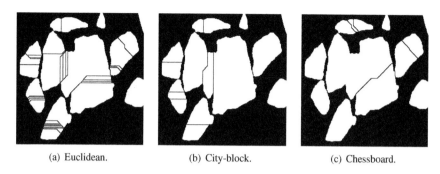

(a) Euclidean.   (b) City-block.   (c) Chessboard.

Figure 5.11   Effects of different distance transform-based watershed segmentations on a binary ice image. None of the methods correctly identifies the cracks

5.1.3   MARKER-CONTROLLED WATERSHED SEGMENTATION

The watershed transform is sensitive to the regional minima in the image, and the over-segmentation is caused by the spurious regional minima determined for an object. Therefore, the segmentation function should first be filtered before computing its watersheds to remove all irrelevant regional minima and hopefully obtain a meaningful result. One of the practical approaches to overcome this problem is to use markers as the set of regional minima, imposing to the segmentation function, to control the number of allowable segmented regions.

The minima imposition technique is a kind of morphological reconstruction for filtering of the image minima. It requires the determination of a set of markers to mark the relevant image objects and their background. Based on the reconstruction by erosion, the imposition of the minima of the input grayscale image g is performed in two steps [146]:

Step 1: The pointwise minimum between the image $g + 1$ and the marker image f is computed: $(g + 1) \wedge f$. Note that two distinct minima to impose may fall within a minimum of g at level 0. It is therefore necessary to consider $(g + 1) \wedge f$ rather than $g \wedge f$. The resulting image is lower or equal to the marker image, and minima are then created at locations corresponding to the markers.

Step 2: The morphological reconstruction by erosion of $(g + 1) \wedge f$ from the marker image f is computed as the modified segmentation function: $g' = R^E_{(g+1) \wedge f}(f)$.

In the marker-controlled watershed segmentation [9], a marker image is an image of the same size as the original image that contains a unique marker (connected component) for each object that belongs to the original image. The size of an object marker can range from a unique pixel to a large connected component of pixels. Typically, large markers perform better than small ones when processing noisy images. Then, the object markers are used as the set of minima to impose to the segmentation function. In other words, the old minima in the segmentation function are replaced by the new ones corresponding to the markers. Finally, the watershed lines are obtained by computing the watersheds of the imposed image [146], and the original image is thereby segmented by superimposing the obtained watershed lines onto it.

To illustrate the process of the marker-controlled watershed segmentation, we will use the binary ice image in Figure 5.8(a) as an example. When applying the city-block distance transform to this binary image, four regional minima consisting of 18 local minimum pixels with different gray scales are found in the resulting city-block distance map, as seen in Figure 5.12(a). As mentioned before, the irrelevant regional minima usually cause over-segmentation when the watershed transform is directly used, with the result that it divides the two connected ice floes into four regions based on those four regional minima. To reduce the number of segmented regions, the operation of dilation with a 5-radius disk structuring element is performed on Figure 5.12(a) to combine the regional minima resulting from the city-block distance transform. This operation results in a marker image containing two connected regions, as seen in Figure 5.12(b). Then we modify the distance map by using morphological reconstruction to impose the regional minima at the locations of the marker. Based on the marker-modified distance map, shown in Figure 5.12(c), the watershed transform separates the two connected ice floes correctly, as seen in Figure 5.12(e).

The use of markers in the watershed segmentation can effectively reduce the over-segmentation and produce a good result. However, the performance of the watershed segmentation depends on how suitable the marker is for the application. The main issue in this approach is in the selection of an appropriate marker. The marker generation procedure just described in the example is simple and easy to implement. However, most of time the connected objects are of irregular shapes and sizes. In such cases, more spurious minima and resulting markers will occur in the distance transform. Those spurious markers might still lead to over-segmentation in the watershed segmentation [150]. Thus, the generation procedures for markers can be much more complex, involving object features, such as size, shape, location, relative distances, texture content, and so on [49]. To determine a suitable marker for a watershed application, this may rely on the knowledge about the expected result, the facts about the image, a priori knowledge, or assumptions about the properties of the object represented in the image. In some applications, it may also require the markers to be defined manually [146]. Therefore, how to choose an appropriate marker is still a challenge for the marker-controlled watershed segmentation.

5.2 COMBINATION OF THE WATERSHED AND NEIGHBORING-REGION MERGING ALGORITHMS

Image segmentation by using the watershed transform is powerful for connected objects segmentation, especially for separation of connected objects with invisible boundaries. However, the over-segmentation with this transform, which is a serious problem that can render the segmented result useless [49], must be overcome. For this reason, we will introduce the neighboring-region merging algorithm in this section to reduce the over-segmentation in the watershed-based ice image segmentation.

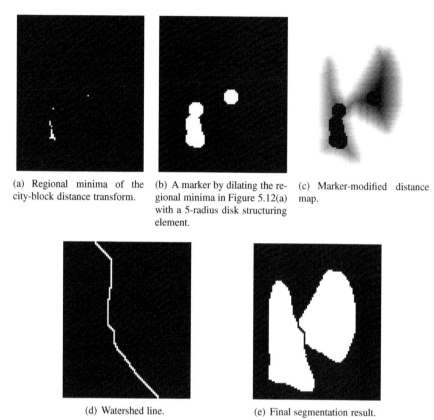

(a) Regional minima of the city-block distance transform.

(b) A marker by dilating the regional minima in Figure 5.12(a) with a 5-radius disk structuring element.

(c) Marker-modified distance map.

(d) Watershed line.

(e) Final segmentation result.

Figure 5.12 Marker-controlled watershed segmentation.

It is assumed for the neighboring-region merging method that each ice floe has a convex boundary and that the junction line between two connected ice floes has at least one concave ending point. Based on this assumption, the junction lines obtained from the watershed-based segmentation are filtered by deleting those that have two convex ending points. Figure 5.13 shows a flow chart of the combination of the watershed segmentation and the neighboring-region merging algorithms. Figure 5.14 and the following steps describe the details of the algorithm:

Step 1: The input ice image is converted into a binary image. This can be done by the Otsu thresholding or k-means clustering methods.

Step 2: The inverse distance map of the binarized ice image is computed, and then the watershed transform is applied to derive the segmented image.

In this step, the distance transform is first computed to the binary image derived from the previous step. Some distance measure must then be selected, such as the city-block, which is a compromise between the three

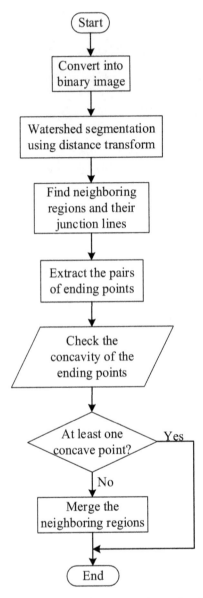

Figure 5.13 Flow chart of the watershed segmentation and neighboring-region merging (Source: Figure from Q. Zhang, R. Skjetne, and B. Su, "Automatic Image Segmentation for Boundary Detection of Apparently Connected Sea-Ice Floes," In *Proceedings of the 22nd International Conference on Port and Ocean Engineering under Arctic Conditions*, Espoo, Finland, 2013).

measures discussed in Section 5.1.2. Then the watershed transform using the 8-neighbor connectivity (the segmented regions are 8-connected neighborhoods and each watershed line forms 4-connected path) is performed on the regional minima in the inverse distance map to obtain a possibly oversegmented ice image, as shown in Figure 5.14(e).

Step 3: We detect the neighboring regions and their junction lines. This is done by the logical AND operation to find the intersection of Figure 5.14(b) with Figure 5.14(d), resulting in Figure 5.14(f).

Step 4: We extract the two ending points of the junction lines. The junction lines yielded by the watershed transform are 4-connected and 1-pixel-thick. Their ending points have only one neighbor in the horizontal or vertical direction. That is, an ending point and its 8 surrounding pixels satisfy one of the forms shown in Figure 5.15(a). Convolving the kernel shown in Figure 5.15(b) with any matrix in Figure 5.15(a) results in a value 3 at the center element of the resulting 3×3 matrix. Thus, by using this kernel to filter Figure 5.14(f), the ending points will be identified at the location where the kernel is centered if the resulting value is equal or larger than 3 (note that if the pixel is an isolated point, the center value becomes larger than 3). The extraction of the ending points can be found by the gray dots in Figure 5.14(f).

Step 5: We then check the concavity of each pair of ending points. If neither of them is a concave point, we merge the corresponding two neighboring regions. For example, the pair of ending points highlighted by gray dots in Figure 5.14(g) are non-concave points. Therefore, their junction line is removed, and the corresponding two neighboring regions are merged, as shown in Figure 5.14(h). This operation is then iterated until all the pairs of ending points are checked. The final segmentation result is seen in Figure 5.14(i).

In this step, a differential chain code method [95] is applied to check the concavity of the two ending points of a proposed junction line. The details of this concave ending point detection is explained in the following section.

5.2.1 CONCAVE DETECTION BY CHAIN CODE

5.2.1.1 Boundary tracing

When executing any chain code algorithm to determine the concavity of the ending points, it requires that the pixels on the boundary of the object are ordered in a clockwise (or counterclockwise) direction. To achieve this, a boundary tracing algorithm is required before using the chain code.

Let O be an object with 1-valued pixels in a binary image, while the background pixels are 0-valued. An idea of the boundary tracing is to assume that a conceptual bug marches around the object's boundary in, for instance, a clockwise direction starting from a 1-valued boundary pixel. When the bug encounters a 1-valued pixel b, it goes back to the previous pixel p (which is marked as a backtracking pixel) from where b is entered. Then, the bug traverses the 8-neighbors of b in clockwise direction until encountering a 1-valued pixel. The algorithm terminates when the bug

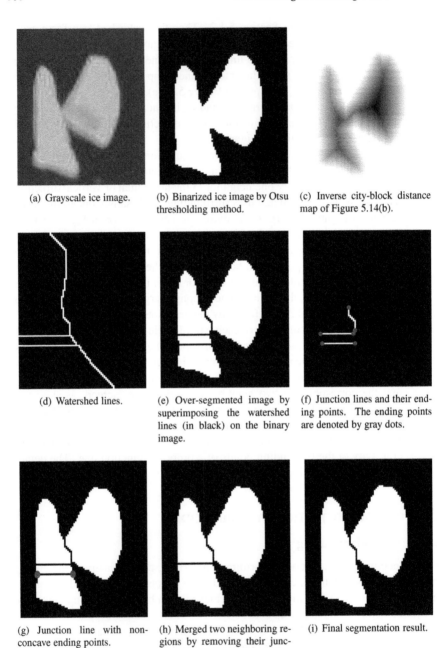

(a) Grayscale ice image.

(b) Binarized ice image by Otsu thresholding method.

(c) Inverse city-block distance map of Figure 5.14(b).

(d) Watershed lines.

(e) Over-segmented image by superimposing the watershed lines (in black) on the binary image.

(f) Junction lines and their ending points. The ending points are denoted by gray dots.

(g) Junction line with non-concave ending points.

(h) Merged two neighboring regions by removing their junction line where the ending points, as highlighted in Figure 5.14(g), are non-concave points.

(i) Final segmentation result.

Figure 5.14   Neighboring-region merging algorithm.

0	1	0
0	1	0
0	0	0

0	0	0
0	1	1
0	0	0

0	0	0
0	1	0
0	1	0

0	0	0
1	1	0
0	0	0

0	1	1
0	1	0
0	0	0

0	0	1
0	1	1
0	0	0

0	0	0
0	1	1
0	0	1

0	0	0
0	1	0
0	1	1

0	0	0
0	1	0
1	1	0

0	0	0
1	1	0
1	0	0

1	0	0
1	1	0
0	0	0

1	1	0
0	1	0
0	0	0

(a) Ending point and its 8 surrounding pixels.

0	-1	0
-1	4	-1
0	-1	0

(b) Kernel for ending
point detection.

Figure 5.15 Ending point detection.

encounters the first two boundary pixels again (the starting pixel and its next boundary pixel) in the same order. While the bug is tracing the boundary, the coordinates of each 1-valued boundary pixel are recorded.

To start the boundary tracing algorithm, let the uppermost, leftmost pixel of the object O (which is 1-valued) be the starting pixel b_0. Let the west neighbor of b_0 be a backtracking pixel p_0 (it is clear that p_0 is always a 0-valued background pixel). Then, the step-by-step procedure of this algorithm is given in the following [48, 49]:

Step 1: Examine the 8-neighbors of b_0 by starting at p_0 and proceeding in a clockwise direction. Let b_1 be the first 1-valued neighbor encountered, and p_1 be the pixel immediately preceding b_1 in the sequence.

Step 2: Set $b = b_1$ and $p = p_1$, and add the position of b_1 in the boundary list.

Step 3: Examine the 8-neighbors of b starting from p in clockwise direction, denoted by $n_1, n_2, \cdots, n_8$. Let n_k be the first 1-valued neighbor encountered.

Step 4: Set $b = n_k$ and $p = n_{k-1}$, and add the position of b in the boundary list.

Step 5: Repeat Steps 3 and 4 until $b = b_0$ and the next boundary pixel found is the same as b_1.

Then the sequence of the b pixels constitutes the set of ordered boundary pixels. The first few steps of this procedure are illustrated in Figure 5.16.

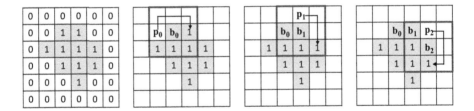

Figure 5.16 First few steps of the boundary tracing algorithm.

In this procedure, p_1 and p are backtracking pixels that are always 0-valued, because b_1 and $b = n_k$ is the first 1-valued pixel found in the clockwise direction. It should be noted that, when examining the 8-neighbors, it does not matter which direction is chosen. However, the chosen direction must be kept consistent throughout the algorithm.

This boundary tracing algorithm detects the exterior boundary of an object in a binary image. Any holes that are present or any objects within holes are ignored. To trace the boundary of a hole within an object, which represent an interior boundary of the object, a hole detection algorithm (e.g., a hole filling algorithm as introduced later in Section 7.1.3) should be used first to extract the hole, and then the boundary tracing algorithm should be applied to the hole (and any other hole found).

5.2.1.2   Differential chain code

As mentioned in Section 2.7, a chain code is used to represent a boundary through a series of specified length straight-line segments in different directions. A change in the chain code direction usually indicates a corner on the boundary. By analyzing the direction changes as we travel in a clockwise direction along the boundary, we can determine and mark the convex and concave corners. However, the chain code has a low accuracy since it only represents 8 directions. Therefore, we use the absolute chain code sum [95] to increase the accuracy.

Assume $C(i)$ and $C(i-1)$ are the chain codes of current node $Node\ i$ and its former node $Node\ i-1$ on a boundary, respectively. Let $R(i)$ denote the relative chain code, and the calculation process for the relative chain code is given by:

$$R(i) = [C(i) - C(i-1) + 8] \mod 8 \tag{5.9a}$$
$$\text{IF } R(i) > 4 \text{ THEN } R(i) = R(i) - 8 \tag{5.9b}$$

The relative chain code is the relationship between $C(i)$ and $C(i-1)$, and it indicates that $Node\ i$ has rotated $R(i) \times 45°$ counter-clockwise (the negative value indicates the clockwise direction) to $Node\ i-1$.

The absolute chain code, denoted by $A(i)$, is the accumulation of the relative chain codes from the starting node to the current $Node\ i$. The starting node of the absolute chain code is set to 0. Therefore, the absolute chain code is expressed as:

$$A(0) = 0 \tag{5.10a}$$
$$A(i) = A(i-1) + R(i) \tag{5.10b}$$

When the total number of boundary nodes is N with the notation from 0 to $N-1$ in clockwise direction, it should be noted that the ending node is denoted as N, which is actually the starting node 0. The difference of the absolute chain codes between the starting node and the ending node is 8, that is,

$$A(0) - A(N) = 8 \tag{5.11}$$

The absolute chain code sum of three sequential nodes is the sum of the absolute chain codes of the current $Node\ i$ and its two former $Node\ i-1$ and $Node\ i-2$, that is,

$$S(i) = A(i) + A(i-1) + A(i-2) \tag{5.12}$$

When calculating this for the two first nodes, the former absolute chain code must be shifted to the end of the sequence and adjusted accordingly, since the boundary is closed, that is,

$$S(0) = A(0) + A(N-1) + A(N-2) + 16 \tag{5.13}$$
$$S(1) = A(1) + A(0) + A(N-1) + 8 \tag{5.14}$$

Similar to Equation 5.11, we get

$$S(0) - S(N) = 24 \tag{5.15}$$

The absolute chain code sum has 24 directions, which is more accurate than the original 8-direction chain code. It can therefore be used to represent the tangent direction of edge points more accurately.

The differential chain code is given by the difference of the absolute chain code sum:

$$D(i) = S(i+3) - S(i) \tag{5.16}$$

When calculating the last 3 nodes, the absolute chain code sum should start at the starting nodes and be adjusted accordingly, that is,

$$D(N-j) = S(N-j+3) - S(N-j) \tag{5.17}$$
$$= S(3-j) - 24 - S(N-j) \tag{5.18}$$

where $j = 1, 2, 3$.

The differential chain code is proportional to the curvature of the edges. It defines the change of angular direction between two neighboring boundary segments:

$$\theta = D(i) \times 15° \tag{5.19}$$

Therefore, a *Node i* can be identified as a concave point if its differential chain code, $D(i)$, is positive when tracing in clockwise direction along the boundary. Considering the roughness of the boundary, the nodes with differential chain code between 3 and 10 are detected as concave points in this research, that is,

Node i is a concave point if $3 \leq D(i) \leq 10$

An example of chain code-based concave detection is shown in Figure 5.17.

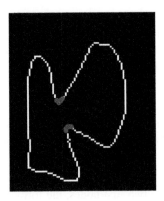

Figure 5.17 Concave detection by chain code, where the detected concave points are highlighted by gray dots.

5.3 EXPERIMENTAL RESULTS AND DISCUSSION

A few sea ice images obtained from the remote sensing expedition of ice conditions, carried out at Ny-Ålesund in early May 2011 [185], are applied in this case study.

It should be noted that the brash ice has been removed manually from the images before segmentation.

Some examples of the processing results are shown in Figure 5.18. As seen, most of the over-segmented lines have been removed, but some unrealistic lines still exist. This is because the applied neighboring-region merging algorithm is based on a very simple assumption, namely that the watershed-segmented line is taken as a correct junction line if it has one or two concave ending points. This assumption is not always correct since a real ice floe is typically not a perfect convex shape. As seen in Figure 5.19(b), the segmented line highlighted by the dark-gray ellipse has a concave ending point, which is just a concave corner of the floe boundary. But this is not a real junction between the two floes. Even if one or both of its ending points are real junction concave corners, the segmented line obtained by the watershed transform is often not a correct junction line. This is illustrated by the segmented line highlighted by the light-gray ellipse in Figure 5.19(b).

Under-segmentation is another problem of the watershed segmentation, and this cannot be improved by the neighboring-region merging algorithm. As shown in Figure 5.19(a), there should be a segmented line in the region highlighted by the gray ellipse, but this is not detected by the algorithm.

Over- and under-segmentation in the boundary detection of the ice floes are the major issues in watershed. Table 5.1 lists the number of over- and under-segmented lines compared to manual inspection in Figure 5.18. It is found that the over-segmented lines are significantly reduced by the neighboring-region merging algorithm, while the under-segmentation is still a problem.

Besides the over- and under-segmentation problems, ambiguously segmented lines are another problem. Figure 5.20 shows an example of this problem. By looking at Figure 5.20(b), it is difficult to say whether the highlighted segmented line really exists. Figure 5.20(c) shows a grayscale image of the same sea ice floes taken under a different reflection condition, featuring the same ice floes with more details. By looking at this image, it seems as if there is a boundary in the highlighted region. However, it is still difficult to identify the exact location of the real crack. Therefore, high-resolution images would definitely provide more accurate ice floe segmentation.

It should also be noted that the real boundary information between the connected floes are actually lost when using the distance transform-based watershed segmentation. This occurs because the distance transform operates on binary images and the watershed transform focuses on the morphological characteristics of the ice floes rather than on the real boundaries. Therefore, the established method is limited by crowded ice floe images in which a massive amount of ice floes are connected to each other and few "holes" or concave regions can be found after binarization (e.g., MIZ images). Thus, the method of this chapter is mainly applicable to ice floe images in which the information of floe junctions is invisible or lost, such as in binary ice images.

Figure 5.18 Connected ice floe segmentation by the established method: (a) grayscale ice images. (b) Binary images with manually identified segmented lines between connected ice floes. (c) Segmented images based on the watershed transform. (d) Segmented images after neighboring-region merging (Source: Figures from Q. Zhang, R. Skjetne, and B. Su, "Automatic Image Segmentation for Boundary Detection of Apparently Connected Sea-Ice Floes," In *Proceedings of the 22nd International Conference on Port and Ocean Engineering under Arctic Conditions*, Espoo, Finland, 2013).

(a) Manually identified segmentation.

(b) Segmentation by the established neighboring-region merge method.

Figure 5.19 Examples of over- and under-segmentation (Source: Figures from Q. Zhang, R. Skjetne, and B. Su, "Automatic Image Segmentation for Boundary Detection of Apparently Connected Sea-Ice Floes," In *Proceedings of the 22nd International Conference on Port and Ocean Engineering under Arctic Conditions*, Espoo, Finland, 2013).

Table 5.1
Number of over- and under-segmented lines in Figure 5.18.

Image	Manual in-spec-tion	Watershed-based segmentation using the distance transform			Watershed segmentation combining with neighboring region merging		
	No. of floes	No. of floes	Over seg-ment	Under seg-ment	No. of floes	Over seg-ment	Under seg-ment
1	5	12	7	0	5	0	0
2	12	16	5	1	11	0	1
3	20	25	7	2	17	0	3
4	38	60	26	5	37	4	6

Source: Table from Q. Zhang, R. Skjetne, and B. Su, "Automatic Image Segmentation for Boundary Detection of Apparently Connected Sea-Ice Floes," In *Proceedings of the 22nd International Conference on Port and Ocean Engineering under Arctic Conditions*, Espoo, Finland, 2013.

(a) Grayscale ice image.

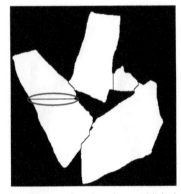

(b) Segmentation by the established method.

(c) Original ice image with more details.

(d) Manually identified segmentation.

Figure 5.20 An example of the ambiguously segmented lines (Source: Figures from Q. Zhang, R. Skjetne, and B. Su, "Automatic Image Segmentation for Boundary Detection of Apparently Connected Sea-Ice Floes," In *Proceedings of the 22nd International Conference on Port and Ocean Engineering under Arctic Conditions*, Espoo, Finland, 2013).

6 GVF Snake-Based Ice Floe Boundary Identification and Ice Image Segmentation

A snake, also called a deformable contour or an active contour [75], is a controlled continuous closed curve that can move and evolve its shape and position under the influence of internal and external forces. The former comes from the curve itself, while the latter consists of external potential forces (typically derived from the image data) together with additional constraint forces. The evolution of the snake stops ideally at some features of interest in the image, where the internal and external forces reach an equilibrium.

The entire problem of finding features, such as edges and object boundaries in an image, is not solved by a snake. It relies on other mechanisms to place the snake somewhere near a desired image feature, such that the energy minimization will carry the snake toward the desired feature.

According to the representation and implementation, the snake models can be classified into two basic types: parametric snake models [75] and geometric snake models [19, 20, 21]. In the parametric snake models, curves (or surfaces in 3-dimensional space) are represented explicitly as parameterized curves (surfaces) in a Lagrangian formulation. For the geometric snake models, on the other hand, these are represented implicitly as a level set of a higher-dimensional function, which evolves according to an Eulerian formulation [175]. Comparing these two types of models, the geometric snake models evolve independently of the parameterization and can automatically handle the topology changes (so that the geometric snake models are able to extract multiple image features simultaneously). However, when the object boundary is weak or has gaps, the geometric snake models may "leak out". The parametric snake models, however, can give several good solutions to this problem [174, 127, 120]. Thus, in this book we consider the parametric snake model only, due to its superior detection capability of weak edges.

To run the parametric snake algorithm, however, the initial contour, which is a starting set of snake points for the evolution, should be placed close to the desired image features. Otherwise, the snake will likely converge to the wrong solution, a local minimum. This happens because the capture range of the external force field, which is the area that drives the convergence of the snake in the direction of the desired feature, is limited. Moreover, when the desired features are object boundaries, the external forces may have no components pulling the snake into concavities along the boundary. Hence, the snake may smoothen the deep boundary concavities.

To overcome the drawbacks of the snake algorithm, many efforts, such as the

109

multiresolution method [83], the distance potential force [30], and the balloon force [29] etc., have been made to try to improve the external forces and convergence properties of the traditional snake. The external forces of the snake can generally be divided into two classes: the static forces, which are computed from the image data and do not change as the snake progresses, and the dynamic forces, which change depending on the snake's evolution [174]. For example, a distance potential force, which is a static external force, uses a distance map as the source of the external force based on the principle that the snake should be attracted to the nearest edge [30]. By defining the external energy function based on the distance map, the capture range of the traditional snake is enlarged. However, the problem of poor convergence into boundary concavities still cannot be solved. In contrast, a dynamic external force, called the balloon force, can either inflate or deflate a closed snake curve by exerting a force that is normal to the snake in outward or inward direction [29]. Balloon forces can pull the snake into a boundary concavity, but this requires to pre-specify the motion direction of the snake by the user or through some prior knowledge. Balloon forces may also have a problem of pulling the snakes to weak edges if they act too strongly. Although a large number of improvements have been proposed for the traditional snake, most of them can solve only one problem while often generating other new problems [174].

Aiming at enlarging the capture range of the external force field as well as enhancing the ability of the snake to progress into deep boundary concavities, a static external force field method called gradient vector flow (GVF) is employed in the snake algorithm[1] [174]. The GVF snake uses a diffusion of the gradient vectors of an edge map as the source of its external forces. This results in a smooth attraction field that is defined in the whole image and spreads the influence of boundary concavities. Therefore, the GVF snake is less sensitive to the initial contour and able to progress into boundary concavities. Due to these advantages, the GVF snake algorithm has become one of the most famous snake algorithms, and it has been widely used in object segmentation and tracking [153, 182, 154, 88, 135, 85].

In this chapter, the GVF snake algorithm is adopted to identify floe boundaries and segment ice images. To avoid user interaction and to reduce the time required to run the GVF snake algorithm, an automatic contour initialization will be proposed and presented based on the distance transform.

6.1 TRADITIONAL PARAMETRIC SNAKE MODEL

Given an arbitrary parametrization of a curve, $\mathbf{c} = \mathbf{c}(t)$, with t as a free variable in the interval $0 \leq t \leq T$, the arc length along the curve is defined as [159]:

$$s = \int_0^t \left| \frac{d\mathbf{c}}{dt} \right| dt \tag{6.1}$$

---

[1]An introduction and MATLAB® codes are available at authors' webpage at http://iacl.ece.jhu.edu/Projects/gvf/.

where the vector $d\mathbf{c}/dt$ is the tangent vector. This is always a unit vector if t is also an arc length parameterized variable.

A traditional snake is represented mathematically as a curve $\mathbf{c}(s) = (x(s), y(s))$ ($s \in [0, 1]$) parameterized by its arc length s, where $x(s)$ and $y(s)$ are the coordinates along the curve, as seen in Figure 6.1. Since we deal only with closed contours, the boundary condition for $\mathbf{c}$ is periodic, that is, $\mathbf{c}(0) = \mathbf{c}(1)$. A snake model represents a compromise between the properties of the snake itself and the properties of the image. The objective is to determine the shape and position of the snake by finding a parameterized curve $\mathbf{c}(s)$ that iteratively minimizes the energy functional [75] given by:

$$\mathbf{E} = \int_0^1 \mathbf{E}_{int}(\mathbf{c}(s)) + \mathbf{E}_{ext}(\mathbf{c}(s)) ds \qquad (6.2)$$

where $\mathbf{E}_{int}$ represents the internal energy of the snake due to stretching or bending, and $\mathbf{E}_{ext}$ represents the external energy of the snake measuring how well the curve fits to the image data. As a result, the snake dynamically updates its shape and location according to the argument that minimizes the energy. This procedure minimization is iterated until it reaches an equilibrium with a stable shape and position that conforms to the feature of interest.

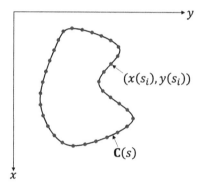

Figure 6.1   A snake curve.

6.1.1   THE ENERGY FUNCTIONALS

The energy functionals are expressed in terms of the snake configuration and of the static image data. These contribute to the snake energy with different purposes.

6.1.1.1   Internal energy

The snake energy $\mathbf{E}_{int}$ models the internal energy of a linear elastic band, corresponding to the internal forces that controls the natural behavior of the snake. It is defined within the contour and designed to keep the snake smooth during the evolution. The internal energy $\mathbf{E}_{\text{int}}$ is influenced by the derivative information of the snake curve, according to:

$$\mathbf{E}_{\text{int}} = \frac{1}{2}\alpha \left| \frac{d\mathbf{c}(s)}{ds} \right|^2 + \frac{1}{2}\beta \left| \frac{d^2\mathbf{c}(s)}{ds^2} \right|^2 \tag{6.3}$$

where α and β are non-negative weighting parameters that control the relative influence of the corresponding energy term at any point s along the snake.

The first-order derivative term is known as the stretching energy or the elastic energy, since large values of this first-order derivative indicate a high rate of change in that region of the snake. This term also corresponds to the continuity of the snake, since it encourages tension in the snake to prevent it from getting stretched. The weighting parameter α controls the contribution of the first-order term to determine the point spacing of the snake. High values for α make the snake attain evenly spaced contour points, whereas low values for α result in a varying spacing of points. Increasing α will increase the tension of the snake, and it tends to eliminate extraneous loops and ripples by reducing the length of the snake. The influence of point spacing of the snake can be removed by setting α to zero. The points along the snake will then become unevenly spaced. An example is given in Figure 6.2 to show how α affects the behavior of the snake.

(a) Initialization of the snake. (b) Termination of the snake by using a large α. (c) Termination of the snake by using a small α.

Figure 6.2 Influence of α on the snake's behavior.

The second-order term is called the bending energy or the curvature energy because of its approximation of the curvature. This term corresponds to the rigidity of the snake. It encourages smoothness and penalizes high curvature to avoid oscillations (such as sharp corners and spikes) of the snake. The weighting parameter β controls the contribution of the second-order term with respect to point variation. High values for β predispose the snake to smooth contours, whereas low values for β mean that the curvature is less minimized and corners can be formed. Increasing β will increase the bending rigidity of the snake and tend to make it smoother and less flexible. The influence of curvature of the snake can be removed by setting β to zero in order to allow the snake to better form corners. An example is given in Figure 6.3 to show how β affects the behavior of the snake.

Both the first-order and the second-order terms correspond to the internal properties of the snake and are independent of the image. Thus, they are called the internal energy terms. These ensure that the movement of a point on the snake curve depends on its neighboring snake points, making the snake less rough or irregular. Both α

(a) Initialization of the snake.   (b) Termination of the snake by using a large β.   (c) Termination of the snake by using a small β.

Figure 6.3 Influence of β on the snake's behavior.

and β can be dynamically dependent on s or they can be constants to control the shape of the snake (they are typically set constant). Decreasing either or both of α and β can be favorable for features with high localized curvature, and setting one or both of them to zero at a point s will allow discontinuities in the snake at s. This is the mutual dependence between α and β [111, 105].

6.1.1.2   External energy

The external energy $\mathbf{E}_{ext}$ corresponds to the external forces that attract the snake to salient features in the image. It is defined on the entire image domain and should take lower values at the desired features [84]. $\mathbf{E}_{ext}$ can be given by:

$$\mathbf{E}_{ext} = \mathbf{E}_{image}(\mathbf{c}(s)) + \mathbf{E}_{con}(\mathbf{c}(s)) \tag{6.4}$$

where $\mathbf{E}_{image}$ represents the image energy, and $\mathbf{E}_{con}$ represents the external constraint energy.

6.1.1.2.1   Image energy

The image energy gives rise to the potential forces that push the snake toward desired low-level image features, such as lines, edges, and terminations. Then the total image energy can be expressed as a weighted combination of these three energy functionals, given by:

$$\mathbf{E}_{image} = \gamma_{line}\mathbf{E}_{line} + \gamma_{edge}\mathbf{E}_{edge} + \gamma_{term}\mathbf{E}_{term} \tag{6.5}$$

where $\mathbf{E}_{line}$, $\mathbf{E}_{edge}$, and $\mathbf{E}_{term}$ denote the energies of line, edge, and terminations, respectively, and their weighting parameters are respective denoted by γ_{line}, γ_{edge}, and γ_{term}.

The line energy $\mathbf{E}_{line}$ can simply be set to the image intensity:

$$\mathbf{E}_{line} = I(x,y) \tag{6.6}$$

where $I(x,y)$ is the image intensity at position (x,y). Then the snake will be attracted either to black (0-valued) or to white (1-valued) lines depending on the sign of γ_{line}:

positive value of γ_{line} for detecting black lines on a white background, while negative value for detecting white lines on a black background.

The edge energy $\mathbf{E}_{edge}$ can be calculated in terms of the image gradient:

$$\mathbf{E}_{edge} = -|\nabla I(x,y)|^2 \tag{6.7}$$

where $\nabla I(x,y) = (I_x, I_y)$ is the image gradient that represents a directional change in the brightness of the image with the gradient angle $\theta = \arctan(I_x/I_y)$, and the negative sign in Equation 6.7 indicates that the sum of the image gradient magnitude over the entire snake needs to be maximized since we want to minimize the energy functional [4]. Then the snake is attracted to edges with large gradient magnitudes. Note that the $\mathbf{E}_{edge}$ becomes very small when the snake get close to an edge.

The termination energy $\mathbf{E}_{term}$ can be measured by the curvature of level contours in a slightly smoothed version of the original image (e.g., by the original grayscale image smoothed by a Gaussian filter) [75]:

$$\begin{aligned}
\mathbf{E}_{term} &= \frac{\partial \theta}{\partial \mathbf{n}_\perp} = \frac{\partial^2 C/\partial \mathbf{n}_\perp^2}{\partial C/\partial \mathbf{n}} \\
&= \frac{C_{yy}C_x^2 - 2C_{xy}C_xC_y + C_{xx}C_y^2}{(C_x^2 + C_y^2)^{3/2}}
\end{aligned} \tag{6.8}$$

where $C(x,y)$ denotes the slightly smoothed image, $\theta = \arctan(C_x/C_y)$ is the gradient angle, $\mathbf{n} = (\cos\theta, \sin\theta)$ and $\mathbf{n}_\perp = (-\sin\theta, \cos\theta)$ are the unit vectors along and perpendicular to the gradient direction, respectively. By combination of $\mathbf{E}_{edge}$ and $\mathbf{E}_{term}$, the snake can be attracted to edges or terminations of line segments and corners. Note, however, that the termination energy $\mathbf{E}_{term}$ is rarely used [111].

6.1.1.2.2 External constraint energy

The constraint energy is a contribution to the external energy in Equation 6.4 that allows higher-level information to control the snakes evolutions. It gives rise to the forces that are responsible for the initialization of the snake near the desired position by some a priori means, such as a user interaction, automatic attentional mechanisms, or some other higher-level computer vision mechanisms. The constraint energy $\mathbf{E}_{con}$ is often not used and set to zero [158].

6.1.1.2.3 External energy for object boundary detection

When object boundaries are the desired features, the external energy is usually taken to be the gradient magnitude of the image, whereas the line and termination energies are set to zero:

$$\mathbf{E}_{ext} = \gamma \mathbf{E}_{edge} = -\gamma |\nabla I(x,y)|^2 \tag{6.9}$$

where is γ is a weighting parameter used to control the importance of the external energy relative to the internal energy.

However, by using the image gradient as the external energy, the capture range will be very limited. This is because the edge information usually causes large gradients (large negative values of $\mathbf{E}_{ext}$) only within a very limited area around object boundaries in the image. Increasing the capture range can be achieved by slightly smoothing the image, for instance, using a Gaussian filter to spatially smooth the image:

$$C(x,y) = G_\sigma(x,y) * I(x,y) \qquad (6.10)$$

where $G_\sigma(x,y) = \frac{1}{2\pi\sigma^2} e^{-(x^2+y^2)/2\sigma^2}$ is a 2-dimensional Gaussian kernel with a standard deviation σ, and $*$ denotes the convolution operator. The external energy is then given by:

$$\mathbf{E}_{ext} = -\gamma|\nabla C(x,y)|^2 = -\gamma|\nabla(G_\sigma(x,y) * I(x,y))|^2 \qquad (6.11)$$

As a result, the image is blurred by a large value of σ, and it will be less accurate and distinct for a snake to localize object boundaries. However, a large σ is often necessary to increase the capture range of the snake.

When the image is a binary image, where the desired features are 1-valued and the background is 0-valued, the appropriate external energies can also be chosen as [29, 30]:

$$\mathbf{E}_{ext} = -\gamma I(x,y) \qquad (6.12)$$

or

$$\mathbf{E}_{ext} = -\gamma C(x,y) = -\gamma G_\sigma(x,y) * I(x,y) \qquad (6.13)$$

The influence of the weighting parameter γ is reduced in the applications where the image data is known to be noisy. However, γ should not be set to zero; otherwise, the snake will evolve under its own energy, without considering any image information, and form a circle that keeps shrinking.

6.1.2 IMPLEMENTATION

According to Equation 6.2, the total snake energy functional is given by:

$$\mathbf{E}(s) = \int_0^1 \frac{1}{2}\alpha \left|\frac{d\mathbf{c}(s)}{ds}\right|^2 + \frac{1}{2}\beta \left|\frac{d^2\mathbf{c}(s)}{ds^2}\right|^2 + \mathbf{E}_{ext}(\mathbf{c}(s))ds \qquad (6.14)$$

To find a contour $\mathbf{c}(s)$ that minimizes the energy functional $\mathbf{E}$, calculus of variations [130] is employed to solve for all snake points in one step to ensure that the snake moves to the best local energy minimum. The minimization problem is then reduced to two independent differential equations, which can be solved numerically [111].

Considering a valid solution $\hat{\mathbf{c}}(s)$ perturbed by a small amount $\varepsilon\delta\mathbf{c}(s)$, for which the energy functional reaches a minimum, by:

$$\frac{d\mathbf{E}(\hat{\mathbf{c}}(s) + \varepsilon\delta\mathbf{c}(s))}{d\varepsilon} = 0 \qquad (6.15)$$

where the perturbation is spatial, affecting the x and y coordinates of a snake point, ε is a small arbitrary scalar, and $\delta\mathbf{c}(s) = (\delta_x(s), \delta_y(s))$ is an arbitrary function of s. Then the perturbed snake solution is:

$$\hat{\mathbf{c}}(s) + \varepsilon\delta\mathbf{c}(s) = (\hat{x}(s) + \varepsilon\delta_x(s), \hat{y}(s) + \varepsilon\delta_y(s)) \qquad (6.16)$$

where $\hat{x}(s)$ and $\hat{y}(s)$ are the x and y coordinates of the snake points at the solution $\hat{\mathbf{c}}(s) = (\hat{x}(s), \hat{y}(s))$, respectively. By substituting the perturbed snake solution into the energy functional given in Equation 6.14, it gives:

$$\mathbf{E}(\hat{\mathbf{c}}(s) + \varepsilon\delta\mathbf{c}(s)) = \int_0^1 \left[\frac{1}{2}\alpha \left| \frac{d(\hat{\mathbf{c}}(s) + \varepsilon\delta\mathbf{c}(s))}{ds} \right|^2 + \frac{1}{2}\beta \left| \frac{d^2(\hat{\mathbf{c}}(s) + \varepsilon\delta\mathbf{c}(s))}{ds^2} \right|^2 \right.$$

$$\left. + \mathbf{E}_{\text{ext}}(\hat{\mathbf{c}}(s) + \varepsilon\delta\mathbf{c}(s)) \right] ds$$

(6.17)

By substitution from Equation 6.16 and expansion of the squared magnitude terms, we obtain:

$$\mathbf{E}(\hat{\mathbf{c}}(s) + \varepsilon\delta\mathbf{c}(s)) = \int_0^1 \left\{ \frac{1}{2}\alpha \left[\left(\frac{d\hat{x}(s)}{ds} \right)^2 + 2\varepsilon\frac{d\hat{x}(s)}{ds}\frac{d\delta_x(s)}{ds} + \left(\varepsilon\frac{d\delta_x(s)}{ds} \right)^2 \right. \right.$$

$$\left. + \left(\frac{d\hat{y}(s)}{ds} \right)^2 + 2\varepsilon\frac{d\hat{y}(s)}{ds}\frac{d\delta_y(s)}{ds} + \left(\varepsilon\frac{d\delta_y(s)}{ds} \right)^2 \right]$$

$$+ \frac{1}{2}\beta \left[\left(\frac{d^2\hat{x}(s)}{ds^2} \right)^2 + 2\varepsilon\frac{d^2\hat{x}(s)}{ds^2}\frac{d^2\delta_x(s)}{ds^2} + \left(\varepsilon\frac{d^2\delta_x(s)}{ds^2} \right)^2 \right.$$

$$\left. + \left(\frac{d^2\hat{y}(s)}{ds^2} \right)^2 + 2\varepsilon\frac{d^2\hat{y}(s)}{ds^2}\frac{d^2\delta_y(s)}{ds^2} + \left(\varepsilon\frac{d^2\delta_y(s)}{ds^2} \right)^2 \right]$$

$$\left. + \mathbf{E}_{\text{ext}}(\hat{\mathbf{c}}(s) + \varepsilon\delta\mathbf{c}(s)) \right\} ds$$

(6.18)

By expanding the external energy $\mathbf{E}_{\text{ext}}$ at the perturbed solution by Taylor series, we derive:

$$\mathbf{E}_{\text{ext}}(\hat{\mathbf{c}}(s) + \varepsilon\delta\mathbf{c}(s)) = \mathbf{E}_{\text{ext}}(\hat{x}(s) + \varepsilon\delta_x(s), \hat{y}(s) + \varepsilon\delta_y(s))$$

$$= \mathbf{E}_{\text{ext}}(\hat{x}(s), \hat{y}(s)) + \varepsilon\delta_x(s) \left. \frac{\partial\mathbf{E}_{\text{ext}}}{\partial x} \right|_{(\hat{x},\hat{y})} + \varepsilon\delta_y(s) \left. \frac{\partial\mathbf{E}_{\text{ext}}}{\partial y} \right|_{(\hat{x},\hat{y})}$$

$$+ O(\varepsilon^2)$$

(6.19)

where $\left. \frac{\partial\mathbf{E}_{\text{ext}}}{\partial x} \right|_{(\hat{x},\hat{y})}$ and $\left. \frac{\partial\mathbf{E}_{\text{ext}}}{\partial y} \right|_{(\hat{x},\hat{y})}$ denote the partial derivatives of the external energy with respect to x and y at point $(\hat{x}, \hat{y})$, respectively. The higher order terms in ε can be ignored since ε is small. Then Equation 6.18 can be reformulated as:

$$\mathbf{E}(\hat{\mathbf{c}}(s) + \varepsilon \delta \mathbf{c}(s)) = \mathbf{E}(\hat{\mathbf{c}}(s))$$

$$+ \varepsilon \int_0^1 \alpha \frac{d\hat{x}(s)}{ds} \frac{d\delta_x(s)}{ds} + \beta \frac{d^2\hat{x}(s)}{ds^2} \frac{d^2\delta_x(s)}{ds^2} + \delta_x(s) \left. \frac{\partial E_{\text{ext}}}{\partial x} \right|_{(\hat{x},\hat{y})} ds \qquad (6.20)$$

$$+ \varepsilon \int_0^1 \alpha \frac{d\hat{y}(s)}{ds} \frac{d\delta_y(s)}{ds} + \beta \frac{d^2\hat{y}(s)}{ds^2} \frac{d^2\delta_y(s)}{ds^2} + \delta_y(s) \left. \frac{\partial E_{\text{ext}}}{\partial y} \right|_{(\hat{x},\hat{y})} ds$$

When the perturbed solution is at a minimum, therefore the two integration terms in Equation 6.20 must be equal to zero, that is:

$$\int_0^1 \alpha \frac{d\hat{x}(s)}{ds} \frac{d\delta_x(s)}{ds} + \beta \frac{d^2\hat{x}(s)}{ds^2} \frac{d^2\delta_x(s)}{ds^2} + \delta_x(s) \left. \frac{\partial E_{\text{ext}}}{\partial x} \right|_{(\hat{x},\hat{y})} ds = 0 \qquad (6.21a)$$

$$\int_0^1 \alpha \frac{d\hat{y}(s)}{ds} \frac{d\delta_y(s)}{ds} + \beta \frac{d^2\hat{y}(s)}{ds^2} \frac{d^2\delta_y(s)}{ds^2} + \delta_y(s) \left. \frac{\partial E_{\text{ext}}}{\partial y} \right|_{(\hat{x},\hat{y})} ds = 0 \qquad (6.21b)$$

By applying the method of integration by parts to the integral in Equation 6.21a, we obtain:

$$\left\{ \alpha \frac{d\hat{x}(s)}{ds} \delta_x(s) \right\} \Bigg|_0^1 - \int_0^1 \frac{d}{ds} \left[\alpha \frac{d\hat{x}(s)}{ds} \right] \delta_x(s) ds$$

$$+ \left\{ \beta \frac{d^2\hat{x}(s)}{ds^2} \frac{d\delta_x(s)}{ds} \right\} \Bigg|_0^1 - \left\{ \frac{d}{ds} \left[\beta \frac{d^2\hat{x}(s)}{ds^2} \right] \delta_x(s) \right\} \Bigg|_0^1 \qquad (6.22)$$

$$+ \int_0^1 \frac{d^2}{ds^2} \left[\beta \frac{d^2\hat{x}(s)}{ds^2} \right] \delta_x(s) ds + \int_0^1 \left. \frac{\partial E_{\text{ext}}}{\partial x} \right|_{(\hat{x},\hat{y})} \delta_x(s) ds = 0$$

Since for a closed contour, $\delta_x(1) - \delta_x(0) = 0$ (and $\delta_y(1) - \delta_y(0) = 0$), the first, third, and fourth terms in Equation 6.22 are equal to zero. Equation 6.22 thus reduces to:

$$\int_0^1 \left\{ -\frac{d}{ds} \left[\alpha \frac{d\hat{x}(s)}{ds} \right] + \frac{d^2}{ds^2} \left[\beta \frac{d^2\hat{x}(s)}{ds^2} \right] + \left. \frac{\partial E_{\text{ext}}}{\partial x} \right|_{(\hat{x},\hat{y})} \right\} \delta_x(s) ds = 0 \qquad (6.23)$$

Since Equation 6.23 holds for all $\delta_x(s)$, the functional derivative must vanish [130], implying:

$$-\frac{d}{ds} \left[\alpha \frac{d\hat{x}(s)}{ds} \right] + \frac{d^2}{ds^2} \left[\beta \frac{d^2\hat{x}(s)}{ds^2} \right] + \left. \frac{\partial E_{\text{ext}}}{\partial x} \right|_{(\hat{x},\hat{y})} = 0 \qquad (6.24)$$

Similarly, Equation 6.21b results in:

$$-\frac{d}{ds} \left[\alpha \frac{d\hat{y}(s)}{ds} \right] + \frac{d^2}{ds^2} \left[\beta \frac{d^2\hat{y}(s)}{ds^2} \right] + \left. \frac{\partial E_{\text{ext}}}{\partial y} \right|_{(\hat{x},\hat{y})} = 0 \qquad (6.25)$$

These two independent partial differential equations are known as the Euler-Lagrange equations. They are derived from the fact that, at the optimum, the derivative of the total snake energy is equal to zero:

$$\nabla E = \nabla E_{\text{int}} + \nabla E_{\text{ext}} = -\frac{\partial}{\partial s} \left[\alpha \frac{\partial \mathbf{c}(s)}{\partial s} \right] + \frac{\partial^2}{\partial s^2} \left[\beta \frac{\partial^2 \mathbf{c}(s)}{\partial s^2} \right] + \nabla E_{\text{ext}}(\mathbf{c}(s)) = 0 \quad (6.26)$$

This can be viewed as a force balance equation [173]:

$$\mathbf{F}_{int} + \mathbf{F}_{ext} = 0 \tag{6.27}$$

where

$$\mathbf{F}_{int} = \frac{\partial}{\partial s}\left[\alpha \frac{\partial \mathbf{c}(s)}{\partial s}\right] - \frac{\partial^2}{\partial s^2}\left[\beta \frac{\partial^2 \mathbf{c}(s)}{\partial s^2}\right] \tag{6.28}$$

is the internal force that discourages stretching and bending, and

$$\mathbf{F}_{ext} = -\nabla \mathbf{E}_{ext}(\mathbf{c}(s)) \tag{6.29}$$

is the external force that pulls the snake toward the desired image features. It will be zero at the maximum of $\mathbf{E}_{ext}$. Thus, it is an influence vector field that points to the desired features in the image, as seen in Figure 6.4. Equation 6.27 gives some insight about the physical behavior of the snake: the snake evolves to the desired image features while trying to achieve an energy-minimal configuration. When it reaches an energy minimum, its internal and external forces reach equilibrium.

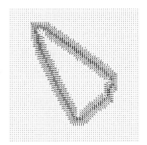

(a) A binary ice floe image.

(b) External energy of Figure 6.4(a), which is the smoothed image gradient by Gaussian filter with $\sigma = 5$.

(c) External force field of Figure 6.4(b).

Figure 6.4 External energy and force field of an ice floe image.

To find a solution of the Euler-Lagrange Equations 6.24 and 6.25, the gradient descent algorithm is typically used as the optimization method. The snake is then made dynamic by treating $\mathbf{c}(s)$ as a function of time t, where the partial derivatives of $\mathbf{c}(s) = (x(s), y(s))$ with respect to t are set equal to the left-hand sides of these partial differential equations, that is [173]:

$$\frac{\partial x(s,t)}{\partial t} = -\frac{\partial}{\partial s}\left[\alpha \frac{\partial x(s,t)}{\partial s}\right] + \frac{\partial^2}{\partial s^2}\left[\beta \frac{\partial^2 x(s,t)}{\partial s^2}\right] + \frac{\partial \mathbf{E}_{ext}}{\partial x}\bigg|_{(x,y)} \tag{6.30a}$$

$$\frac{\partial y(s,t)}{\partial t} = -\frac{\partial}{\partial s}\left[\alpha \frac{\partial y(s,t)}{\partial s}\right] + \frac{\partial^2}{\partial s^2}\left[\beta \frac{\partial^2 y(s,t)}{\partial s^2}\right] + \frac{\partial \mathbf{E}_{ext}}{\partial y}\bigg|_{(x,y)} \tag{6.30b}$$

where $\frac{\partial x(s,t)}{\partial t}$ and $\frac{\partial y(s,t)}{\partial t}$ can be solved in parallel. When the solution $\mathbf{c}(s,t) = (x(s,t), y(s,t))$ becomes stationary, the terms $\frac{\partial x(s,t)}{\partial t}$ and $\frac{\partial y(s,t)}{\partial t}$ tend to zero. Then the energy **E** reaches a minimum, and the snake converges towards the target image features. A numerical solution can be found by discretizing the Equations 6.30a and 6.30b, and then iteratively adjust the discrete system until convergence is achieved [75].

The snake curve $\mathbf{c}(s)$ is practically discretized into a finite number of points $\mathbf{c}(s_i) = (x(s_i), y(s_i)) = (x(ih), y(ih))$ with $i \in \{1, 2, \cdots, N\}$, where each point along the curve is equally spaced by a suitable small arc length (step size in spacing) h. The derivatives through the curve can thereby be approximated with finite differences [140]:

$$\frac{\partial \mathbf{c}(s_i)}{\partial s} \approx \frac{\mathbf{c}(s_i) - \mathbf{c}(s_{i-1})}{h} \tag{6.31}$$

$$\frac{\partial^2 \mathbf{c}(s_i)}{\partial s^2} \approx \frac{\partial}{\partial s}\left(\frac{\partial \mathbf{c}(s_i)}{\partial s}\right) = \frac{\mathbf{c}(s_{i+1}) - 2\mathbf{c}(s_i) + \mathbf{c}(s_{i-1})}{h^2} \tag{6.32}$$

where $h = |s_i - s_{i-1}|$, and an optimization task is needed to estimate h.

Then, the Euler-Lagrange Equation 6.24 can be rewritten as:

$$
\begin{aligned}
&-\frac{1}{h}\left\{\alpha_{s_{i+1}}\frac{x(s_{i+1}) - x(s_i)}{h} - \alpha_{s_i}\frac{x(s_i) - x(s_{i-1})}{h}\right\} \\
&+\frac{1}{h^2}\left\{\beta_{s_{i+1}}\frac{x(s_{i+2}) - 2x(s_{i+1}) + x(s_i)}{h^2} - 2\beta_{s_i}\frac{x(s_{i+1}) - 2x(s_i) + x(s_{i-1})}{h^2}\right. \\
&\left.+\beta_{s_{i-1}}\frac{x(s_i) - 2x(s_{i-1}) + x(s_{i-2})}{h^2}\right\} + \left.\frac{\partial \mathbf{E}_{ext}}{\partial x}\right|_{(x(s_i), y(s_i))} = 0
\end{aligned}
\tag{6.33}
$$

Note again the periodic boundary condition for the closed contour. The index of the points $\mathbf{c}(s_i)$ is cyclic, that is, $s_0 = s_N$, $s_1 = s_{N+1}$, $s_2 = s_{N+2}$, and $s_{-1} = s_{N-1}$. For the image grid constraint, the value of $\frac{\partial \mathbf{E}_{ext}}{\partial x}$ at any location $(x(s_i), y(s_i))$ can be calculated numerically by a bilinear interpolation of the values of $\frac{\partial \mathbf{E}_{ext}}{\partial x}$ at the image grid points near $(x(s_i), y(s_i))$ [173].

Equation 6.33 holds for all s_i. By collecting the coefficients of different points, this equation can be expressed as:

$$a_i x(s_{i-2}) + b_i x(s_{i-1}) + c_i x(s_i) + d_i x(s_{i+1}) + e_i x(s_{i+2}) + \frac{\partial \mathbf{E}_{ext}(s_i)}{\partial x} = 0 \tag{6.34}$$

where

$$a_i = \frac{\beta_{s_{i-1}}}{h^4}$$

$$b_i = -\frac{2(\beta_{s_i} + \beta_{s_{i-1}})}{h^4} - \frac{\alpha_{s_i}}{h^2}$$

$$c_i = \frac{\beta_{s_{i+1}} + 4\beta_{s_i} + \beta_{s_{i-1}}}{h^4} + \frac{\alpha_{s_{i+1}} + \alpha_{s_i}}{h^2}$$

$$d_i = -\frac{2(\beta_{s_{i+1}} + \beta_{s_i})}{h^4} - \frac{\alpha_{s_{i+1}}}{h^2}$$

$$e_i = \frac{\beta_{s_{i+1}}}{h^4}$$

Note, the first- and second-order derivatives through the curve can be simply approximated as:

$$\frac{\partial \mathbf{c}(s_i)}{\partial s} \approx \mathbf{c}(s_i) - \mathbf{c}(s_{i-1}) \tag{6.35}$$

$$\frac{\partial^2 \mathbf{c}(s_i)}{\partial s^2} \approx \mathbf{c}(s_{i+1}) - 2\mathbf{c}(s_i) + \mathbf{c}(s_{i-1}) \tag{6.36}$$

The coefficients in Equation 6.34 then become:

$$a_i = \beta_{s_{i-1}}$$
$$b_i = -2(\beta_{s_i} + \beta_{s_{i-1}}) - \alpha_{s_i}$$
$$c_i = \beta_{s_{i+1}} + 4\beta_{s_i} + \beta_{s_{i-1}} + \alpha_{s_{i+1}} + \alpha_{s_i}$$
$$d_i = -2(\beta_{s_{i+1}} + \beta_{s_i}) - \alpha_{s_{i+1}}$$
$$e_i = \beta_{s_{i+1}}$$

Thus, Equation 6.33 is in the form of a linear set of equations:

$$\mathbf{A}\mathbf{x} + \mathbf{f_x} = 0 \tag{6.37}$$

where

$$\mathbf{A} = \begin{bmatrix} c_1 & d_1 & e_1 & 0 & \cdots & 0 & a_1 & b_1 \\ b_2 & c_2 & d_2 & e_2 & 0 & \cdots & 0 & a_2 \\ a_3 & b_3 & c_3 & d_3 & e_3 & 0 & \cdots & 0 \\ 0 & a_4 & b_4 & c_4 & d_4 & e_4 & 0 & \cdots \\ \vdots & 0 & \ddots & \ddots & \ddots & \ddots & \ddots & 0 \\ 0 & \cdots & 0 & a_{N-2} & b_{N-2} & c_{N-2} & d_{N-2} & e_{N-2} \\ e_{N-1} & 0 & \cdots & 0 & a_{N-1} & b_{N-1} & c_{N-1} & d_{N-1} \\ d_N & e_N & 0 & \cdots & 0 & a_N & b_N & c_N \end{bmatrix}$$

$$\mathbf{x} = \begin{bmatrix} x(s_1) & x(s_2) & x(s_3) & \cdots & x(s_{N-2}) & x(s_{N-1}) & x(s_N) \end{bmatrix}^T$$

$$\mathbf{f_x} = \begin{bmatrix} \frac{\partial E_{ext}(s_1)}{\partial x} & \frac{\partial E_{ext}(s_2)}{\partial x} & \frac{\partial E_{ext}(s_3)}{\partial x} & \cdots & \frac{\partial E_{ext}(s_{N-2})}{\partial x} & \frac{\partial E_{ext}(s_{N-1})}{\partial x} & \frac{\partial E_{ext}(s_N)}{\partial x} \end{bmatrix}^T$$

Similarly, the Euler-Lagrange Equation 6.25 can be written in a matrix form:

$$\mathbf{A}\mathbf{y} + \mathbf{f_y} = 0 \tag{6.38}$$

where

$$\mathbf{y} = \begin{bmatrix} y(s_1) & y(s_2) & y(s_3) & \cdots & y(s_{N-2}) & y(s_{N-1}) & y(s_N) \end{bmatrix}^T$$

$$\mathbf{f_y} = \begin{bmatrix} \frac{\partial E_{ext}(s_1)}{\partial y} & \frac{\partial E_{ext}(s_2)}{\partial y} & \frac{\partial E_{ext}(s_3)}{\partial y} & \cdots & \frac{\partial E_{ext}(s_{N-2})}{\partial y} & \frac{\partial E_{ext}(s_{N-1})}{\partial y} & \frac{\partial E_{ext}(s_N)}{\partial y} \end{bmatrix}^T$$

The discrete Euler-Lagrange Equations 6.37 and 6.38 can be solved analytically. In practice, however, the solution has to be calculated iteratively. Similar to the manner of solving the continuous Euler-Lagrange Equations 6.24 and 6.25, given by Equations 6.30a and 6.30b, an artificial time factor t is introduced to the discrete Euler-Lagrange Equations 6.37 and 6.38. As long as the solution has not converged, the time derivatives of the left-hand sides of the Euler-Lagrange equations are not zero. Thus, the right-hand sides of the equations do not vanish.

To employ an iterative scheme, the difference between two successive steps of the iteration can be considered, that is: the right-hand sides of the discrete Euler-Lagrange Equations 6.37 and 6.38 are set equal to the product of a step size and the negative time derivatives of the left-hand sides. An explicit Euler method is then used for the external forces, by assuming that $\mathbf{f_x}$ and $\mathbf{f_y}$ are constant during a time step, whereas an implicit Euler step is taken for the internal forces that are completely specified by the matrix $\mathbf{A}$. This results in the following iterative process:

$$\mathbf{A}\mathbf{x}^{t+1} + \mathbf{f_x}\left(\mathbf{x}^t, \mathbf{y}^t\right) = -\lambda\left(\mathbf{x}^{t+1} - \mathbf{x}^t\right) \tag{6.39a}$$

$$\mathbf{A}\mathbf{y}^{t+1} + \mathbf{f_y}\left(\mathbf{x}^t, \mathbf{y}^t\right) = -\lambda\left(\mathbf{y}^{t+1} - \mathbf{y}^t\right) \tag{6.39b}$$

where the superscripts denote the time index, and λ corresponds to the step size in time controlling the rate of the snake evolution. Low values for λ lead to fast evolution of the snake, but this might make the snake pass over desired image features in a single step. High values for λ lead to slow evolution of the snake, making the evolution process rather tedious. A very rigid snake can be solved with large step sizes because of the implicit Euler steps for the internal forces. However, larger external forces require much smaller step sizes due to the explicit Euler steps with respect to the external forces [75]. Hence, the appropriate choice for λ is a compromise.

Equations 6.39a and 6.39b can be solved by matrix inversion, leading to the following iteration step:

$$\mathbf{x}^{t+1} = (\mathbf{A} + \lambda\mathbf{I})^{-1}\left(\lambda\mathbf{x}^t - \mathbf{f_x}\left(\mathbf{x}^t, \mathbf{y}^t\right)\right) \tag{6.40a}$$

$$\mathbf{y}^{t+1} = (\mathbf{A} + \lambda\mathbf{I})^{-1}\left(\lambda\mathbf{y}^t - \mathbf{f_y}\left(\mathbf{x}^t, \mathbf{y}^t\right)\right) \tag{6.40b}$$

where $\mathbf{I}$ is the identity matrix. The matrix $\mathbf{A} + \lambda\mathbf{I}$ is pentadiagonal. It does not change if α, β, and λ are all constants; otherwise, it has to be updated at each iteration step. The inverse $(\mathbf{A} + \lambda\mathbf{I})^{-1}$ can be calculated by an **LU** decomposition in $O(N)$ time (the decomposition needs only to be performed once during the evolution of the snake if α, β, and λ are constants). Hence, Equations 6.40a and 6.40b give a rapid solution for the snake to find an energy minimum.

An example is given in Figure 6.5 to show this iterative procedure of the traditional snake for boundary detection. The dark-gray curve is the initial contour, the light-gray curves are several iterations of the snake algorithm, and the black curve is the final detected boundary[2].

It should be noted that the energy function has many local minima, and this iterative procedure will typically lead the snake to the closest local minimum. Hence, it cannot guarantee that the snake will find the desired image feature. Initializing the snake contour sufficiently close to the desired feature will remedy this problem.

---

[2]Unless otherwise specified, the parameters $\alpha = 0.05$ and $\beta = 0.0$ are used for all snakes in this book.

Figure 6.5 The iterative procedure of a traditional snake for boundary detection. The dark-gray curve is the initial contour, the light-gray curves are iterative runs of the snake algorithm, and the black curve is the final detected boundary

6.1.3 LIMITATIONS

The traditional snake algorithm can solve a number of image segmentation problems effectively, particularly in detection of weak boundaries when using an edge map of the image (which is an image gradient) as the external energy. Equation 6.29 implies that the gradient of the edge map is used as the external forces in the traditional snake model. The edge map $f(x,y)$ derived from an image $I(x,y)$ has higher values near the image edges, and its gradient ∇f has the following three important properties:

1. The vectors of ∇f point toward the image edges. On the edges they are normal vectors.
2. The vectors of ∇f generally have large magnitudes only in the immediate vicinity of the edges.
3. ∇f is almost zero in the homogeneous regions of the image, where the intensity values of the image are nearly constant.

It follows that the external forces $\mathbf{F}_{ext}$ have large values near the boundaries and small or zero values in the homogeneous regions, and this results in a limited capture range of the external force fields of the traditional snake algorithm, as seen in Figure 6.4(c). Consequently, it is difficult for a snake to converge in regions of low variations in intensity. It is therefore sensitive to the initial contour, and the initial contour should be somewhat close to the true boundary. Otherwise, the curve will evolve mainly (or only) under its own internal forces and likely converge to an incorrect result. This is seen in Figure 6.6 where the initial contour is far away from the desired boundary. Additionally, the traditional snake sometimes fails to progress into deep boundary concavities. This is illustrated in Figure 6.7(a), where there is a failure of convergence of a traditional snake into the concavity of a U-shaped object. Figure 6.7(c) shows a close look at the external force field within the concave region. The external forces are correctly shown to point toward the object boundary. However, the external forces in the channel region that leads into the concavity are purely horizontal, and no force is therefore present to pull the snake down into the concavity. This is typical as to why the traditional snake struggles to progress into deep boundary concavities.

Figure 6.6 Incorrect convergence of a traditional snake in the case where the initial contour is far away from the desired boundary. The dark-gray curve is the initial contour, the light-gray curves are iterative runs of the snake algorithm, and the black curve is the final detected boundary

6.2 GRADIENT VECTOR FLOW (GVF) SNAKE

To overcome the limitations of the traditional snake, the GVF snake replaces the external force $-\nabla \mathbf{E}_{\text{ext}}$ in the Euler-Lagrange Equation 6.26 with a GVF field $\mathbf{v}(x,y)$. The intention is to increase the capture range of the external force fields from the boundary regions to the homogeneous regions, and enable the external forces to point into deep concavities of object boundaries [174].

A GVF field $\mathbf{v}(x,y) = (u(x,y), v(x,y))$ is computed as a spatial diffusion of the gradient vectors of an edge map, derived from the image by minimizing the energy functional:

$$
\begin{aligned}
\mathcal{E} &= \iint \mu |\nabla \mathbf{v}|^2 + |\nabla f|^2 |\mathbf{v} - \nabla f|^2 \, dx dy \\
&= \iint \mu \left(u_x^2 + u_y^2 + v_x^2 + v_y^2 \right) + \left(f_x^2 + f_y^2 \right) \left[(u - f_x)^2 + (v - f_y)^2 \right] dx dy
\end{aligned}
\tag{6.41}
$$

where u_x, u_y, v_x, v_y are the derivatives of the vector field in the x and y directions of the image, respectively, and μ is a regularization parameter that controls the balance between the first and second order terms in the integrand. The edge map f, which is an image gradient taking a larger value on image edges, can be derived by using any edge detector. If the features of interest are other image features rather than edges, f can be redefined to be larger at the desired features.

In this energy functional, $|\nabla f|$ becomes large when getting closer to the object boundaries, in which case the second term dominates the integrand. Consequently, $\mathcal{E}$ is minimized by setting $\mathbf{v}$ very close to or equal to ∇f, and the GVF snake behaves very similarly to the traditional snake. On the other hand, since $|\nabla f|$ is small in the homogeneous regions of the image, the first term, which is the smoothing term enforcing a slow variation of $\mathbf{v}$ toward the edges, dominates the integrand. By the minimization of $\mathcal{E}$, this ensures that the external force field varies slowly and still acts

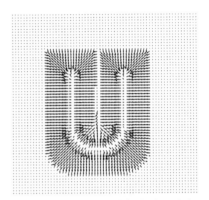

(a) The iterations of a traditional snake to a U-shaped object.

(b) External force field of the U-shaped object image by using the Gaussian smoothed image gradient with $\sigma = 4$.

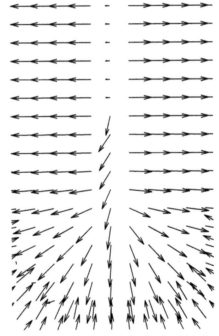

(c) A close look at the external force field in the boundary concavity region.

Figure 6.7 The progression of a traditional snake to a U-shaped feature. The dark-gray curve is the initial contour, the light-gray curves are iterative runs of the snake algorithm, and the black curve is the final detected boundary.

in the homogeneous regions. The parameter μ should be set according to the level of noise present in the image. Usually the more noise contained in the image, the higher value should be selected for μ, resulting in a slowly-varying gradient vector field that is not influenced by spurious gradients in the image[3]. Briefly speaking, the energy functional produces a vector field $\mathbf{v}$ that is nearly equal to the gradient of the edge map around object boundary regions and vary slowly toward the edges in the homogeneous regions.

The calculation of the GVF field is not straightforward when compared with the external force field in the traditional snake model (e.g., the gradients of Equations 6.9, 6.11, 6.12, and 6.13). However, it can be determined by variational minimization and discretization, in a similar way to the snake energy functional minimization [4, 172].

We assume $\hat{\mathbf{v}}(u,v) = (\hat{u}(x,y), \hat{v}(x,y))$ is an admissible GVF solution. By adding small perturbations functions $\varepsilon p(x,y)$ and $\varepsilon q(x,y)$ (where ε is a small arbitrary scalar), respectively, to $\hat{u}(x,y)$ and $\hat{v}(x,y)$, the GVF energy functional becomes:

$$
\begin{aligned}
\mathcal{E}\left(\hat{u}+\varepsilon p, \hat{v}+\varepsilon q\right) = \iint & \left\{ \mu \left[(\hat{u}_x+\varepsilon p_x)^2 + (\hat{u}_y+\varepsilon p_y)^2 + (\hat{v}_x+\varepsilon q_x)^2 + (\hat{v}_y+\varepsilon q_y)^2 \right] \right. \\
& \left. + (f_x^2+f_y^2)\left[(\hat{u}+\varepsilon p - f_x)^2 + (\hat{v}+\varepsilon q - f_y)^2 \right] \right\} dxdy \\
= \iint & \left\{ \mu \left[(\hat{u}_x^2 + 2\varepsilon\hat{u}_x p_x + \varepsilon^2 p_x^2) + (\hat{u}_y^2 + 2\varepsilon\hat{u}_y p_y + \varepsilon^2 p_y^2) \right] \right. \\
& + \mu \left[(\hat{v}_x^2 + 2\varepsilon\hat{v}_x q_x + \varepsilon^2 q_x^2) + (\hat{v}_y^2 + 2\varepsilon\hat{v}_y q_y + \varepsilon^2 q_y^2) \right] \\
& + (f_x^2+f_y^2)\left[(\hat{u}-f_x)^2 + 2\varepsilon(\hat{u}-f_x)p + \varepsilon^2 p^2 \right] \\
& \left. + (f_x^2+f_y^2)\left[(\hat{v}-f_y)^2 + 2\varepsilon(\hat{v}-f_y)q + \varepsilon^2 q^2 \right] \right\} dxdy
\end{aligned}
$$

$$(6.42)$$

Ignoring higher order terms in ε (since ε is small) and rewriting Equation 6.42 gives:

$$
\begin{aligned}
\mathcal{E}\left(\hat{u}+\varepsilon p, \hat{v}+\varepsilon q\right) = & \; \mathcal{E}\left(\hat{u}, \hat{v}\right) \\
& + 2\varepsilon \iint \left[\mu(\hat{u}_x p_x + \hat{u}_y p_y) + (f_x^2+f_y^2)(\hat{u}-f_x)p \right] dxdy \\
& + 2\varepsilon \iint \left[\mu(\hat{v}_x q_x + \hat{v}_y q_y) + (f_x^2+f_y^2)(\hat{v}-f_y)q \right] dxdy
\end{aligned}
$$

$$(6.43)$$

Since $(\hat{u}+\varepsilon p, \hat{v}+\varepsilon q)$ is the perturbed GVF solution, the GVF energy functional derivative is at stationary, that is:

$$
\frac{d\mathcal{E}\left(\hat{u}+\varepsilon p, \hat{v}+\varepsilon q\right)}{d\varepsilon} = 0
$$

$$(6.44)$$

---

[3]Unless otherwise specified, the parameter μ is set to 0.1 for all GVF snakes in this book.

It follows that the two integration terms in Equation 6.43 must be identically zero:

$$\iint \left[\mu \left(\hat{u}_x p_x + \hat{u}_y p_y \right) + \left(f_x^2 + f_y^2 \right) \left(\hat{u} - f_x \right) p \right] dxdy = 0 \qquad (6.45a)$$

$$\iint \left[\mu \left(\hat{v}_x q_x + \hat{v}_y q_y \right) + \left(f_x^2 + f_y^2 \right) \left(\hat{v} - f_y \right) q \right] dxdy = 0 \qquad (6.45b)$$

Applying integration by parts to the integral in Equation 6.45a,

$$\iint \left[\mu \left(\hat{u}_x p_x + \hat{u}_y p_y \right) + \left(f_x^2 + f_y^2 \right) \left(\hat{u} - f_x \right) p \right] dxdy$$

$$= \mu \left[\int \left(p \hat{u}_x - \int p \hat{u}_{xx} dx \right) dy + \int \left(p \hat{u}_y - \int p \hat{u}_{yy} dy \right) dx \right]$$

$$+ \iint \left(f_x^2 + f_y^2 \right) \left(\hat{u} - f_x \right) p dxdy$$

$$= \mu \left[\int_{\partial \Omega} p \left(\hat{u}_y dx + \hat{u}_x dy \right) - \iint_{\Omega} p \left(\hat{u}_{xx} + \hat{u}_{yy} \right) dxdy \right] + \iint_{\Omega} \left(f_x^2 + f_y^2 \right) \left(\hat{u} - f_x \right) p dxdy$$

$$= \mu \left[\int_{\partial \Omega} p \left(\nabla \hat{u} \cdot d\sigma \right) - \iint_{\Omega} p \left(\hat{u}_{xx} + \hat{u}_{yy} \right) dxdy \right] + \iint_{\Omega} \left(f_x^2 + f_y^2 \right) \left(\hat{u} - f_x \right) p dxdy$$

$$= 0$$

$$(6.46)$$

where Ω is the image domain, $\partial \Omega$ is the boundary of Ω, and $d\sigma$ represents a small element on the boundary $\partial \Omega$.

The natural boundary condition implies that $\nabla \hat{u} \cdot d\sigma = 0$ on boundary $\partial \Omega$. Thus, Equation 6.46 reduces to:

$$- \iint_{\Omega} \left[\mu \left(\hat{u}_{xx} + \hat{u}_{yy} \right) - \left(f_x^2 + f_y^2 \right) \left(\hat{u} - f_x \right) \right] p dxdy = 0 \qquad (6.47)$$

Equation 6.47 shall hold for all p, thus:

$$\mu \left(\hat{u}_{xx} + \hat{u}_{yy} \right) - \left(f_x^2 + f_y^2 \right) \left(\hat{u} - f_x \right) = 0 \qquad (6.48)$$

Similarly, Equation 6.45b is derived into:

$$\mu \left(\hat{v}_{xx} + \hat{v}_{yy} \right) - \left(f_x^2 + f_y^2 \right) \left(\hat{v} - f_y \right) = 0 \qquad (6.49)$$

Therefore, the GVF field can be calculated by optimizing $u(x, y)$ and $v(x, y)$ separately via the following Euler-Lagrange equations:

$$\mu \nabla^2 u - (u - f_x) \left(f_x^2 + f_y^2 \right) = 0 \qquad (6.50a)$$

$$\mu \nabla^2 v - (v - f_y) \left(f_x^2 + f_y^2 \right) = 0 \qquad (6.50b)$$

where ∇^2 is the Laplacian operator. In the homogeneous regions where $I(x, y)$ is nearly constant, the gradients of $f(x, y)$ are equal to zero. Thus the second terms in these equations are zero, and the vector field u and v are determined by Laplace's equations in such regions. This makes the GVF field interpolated from object boundaries with a reflection of a kind of competition among the boundary vectors. The result is that the GVF field vectors can point into boundary concavities.

A solution to Equations 6.50a and 6.50b can be obtained by introducing a time variable t and finding the steady-state solution of the following partial differential equations:

$$\frac{\partial u(x,y,t)}{\partial t} = \mu \nabla^2 u(x,y,t) - (u(x,y,t) - f_x(x,y))\left(f_x^2(x,y) + f_y^2(x,y)\right) \quad (6.51a)$$

$$\frac{\partial v(x,y,t)}{\partial t} = \mu \nabla^2 v(x,y,t) - (v(x,y,t) - f_y(x,y))\left(f_x^2(x,y) + f_y^2(x,y)\right) \quad (6.51b)$$

These two equations are known as the generalized diffusion equations that diffuse the gradient of the edge map in regions distant from the boundaries. To obtain their solutions, an explicit difference scheme is adopted, and the partial derivatives in the equations are approximated numerically by the method of finite differences:

$$\frac{\partial u(x,y,t)}{\partial t} = \frac{1}{\Delta t}\left(u^{t+1}(x,y) - u^t(x,y)\right) \quad (6.52a)$$

$$\frac{\partial v(x,y,t)}{\partial t} = \frac{1}{\Delta t}\left(v^{t+1}(x,y) - v^t(x,y)\right) \quad (6.52b)$$

$$\nabla^2 u(x,y,t) = \frac{1}{\Delta x \Delta y}\left(u^t(x+1,y) + u^t(x,y+1)\right.$$
$$\left. + u^t(x-1,y) + u^t(x,y-1) - 4u^t(x,y)\right) \quad (6.52c)$$

$$\nabla^2 v(x,y,t) = \frac{1}{\Delta x \Delta y}\left(v^t(x+1,y) + v^t(x,y+1)\right.$$
$$\left. + v^t(x-1,y) + v^t(x,y-1) - 4v^t(x,y)\right) \quad (6.52d)$$

where Δx and Δy are the spacing between the pixels in the x and y directions, respectively, and Δt is the time-step for each iteration.

By substituting these numerical approximations of the partial derivatives into Equations 6.51a and 6.51b, the numerical solution to the GVF is determined when the following iterative process reaches the steady-state:

$$u^{t+1}(x,y) = \left[1 - \left(f_x^2(x,y) + f_y^2(x,y)\right)\Delta t\right]u^t(x,y)$$
$$+ r\left(u^t(x+1,y) + u^t(x,y+1) + u^t(x-1,y) + u^t(x,y-1) - 4u^t(x,y)\right)$$
$$+ f_x(x,y)\left(f_x^2(x,y) + f_y^2(x,y)\right)\Delta t$$

$$(6.53a)$$

$$v^{t+1}(x,y) = \left[1 - \left(f_x^2(x,y) + f_y^2(x,y)\right)\Delta t\right]v^t(x,y)$$
$$+ r\left(v^t(x+1,y) + v^t(x,y+1) + v^t(x-1,y) + v^t(x,y-1) - 4v^t(x,y)\right)$$
$$+ f_y(x,y)\left(f_x^2(x,y) + f_y^2(x,y)\right)\Delta t$$

$$(6.53b)$$

where

$$r = \frac{\mu \Delta t}{\Delta x \Delta y} \tag{6.54}$$

For stability of the numerical solution to the partial differential equations 6.53a and 6.53b, the step-size constraint $r \leq \frac{1}{4}$ should be maintained according to the Courant-Friedrichs-Lewy (CFL) condition [35]. Thus, to guarantee the convergence of the GVF, the time-step Δt should satisfy:

$$\Delta t \leq \frac{\Delta x \Delta y}{4\mu} \tag{6.55}$$

This implies that when Δx and Δy are large, in which case the image is coarser, the time-step can be larger and the convergence becomes faster; when μ is large, in which case the GVF field is desired to be smoother, this requires a small time-step and the convergence becomes slower.

The GVF field needs to be determined only once, prior to the snake optimization. Once this is done, the external force (the partial derivatives of the external energy) in the traditional snake dynamic Equations 6.30a and 6.30b is replaced with $-\mathbf{v}(x,y)$, yielding:

$$\frac{\partial x(s,t)}{\partial t} = -\frac{\partial}{\partial s}\left[\alpha \frac{\partial x(s,t)}{\partial s}\right] + \frac{\partial^2}{\partial s^2}\left[\beta \frac{\partial^2 x(s,t)}{\partial s^2}\right] - u(x,y) \tag{6.56a}$$

$$\frac{\partial y(s,t)}{\partial t} = -\frac{\partial}{\partial s}\left[\alpha \frac{\partial y(s,t)}{\partial s}\right] + \frac{\partial^2}{\partial s^2}\left[\beta \frac{\partial^2 y(s,t)}{\partial s^2}\right] - v(x,y) \tag{6.56b}$$

These two dynamic equations are the evolution equations for the GVF snake, and they can be optimized numerically by discretization and iteration in the same way as the traditional snake.

Compared to the external force fields in the traditional snake model, having only f_x and f_y, the new vector field with u and v in the GVF are derived by using an iterative method to find a solution for f_x and f_y. The result is that the capture range is effectively enlarged and the initial contour no longer needs to be as close to the true boundaries, as seen in Figure 6.8.

Moreover, the GVF snake has a better ability of convergence into concave regions. For example, taking a close look at the GVF force field in the concavity shown in Figure 6.9(c), the GVF forces have downward components in the channel region that points into the concavity. So the snake will now be pulled into the U-shaped object of Figure 6.9(a).

6.3 CONTOURS INITIALIZATION FOR APPLYING THE GVF SNAKE ALGORITHM IN ICE FLOE BOUNDARY IDENTIFICATION

The GVF snake algorithm is able to detect the weak connections between floes and ensure that the detected boundaries are closed. As illustrated in Figure 6.10, given an initial contour (dark-gray curve), the snake finds the floe boundary (black curve) after

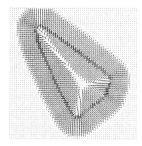 

(a) GVF force field of Figure 6.4(a) computed from the edge map shown in Figure 6.4(b).

(b) The iterations of a GVF snake for boundary detection.

Figure 6.8 GVF force field of Figure 6.4(a) and the iterative procedure of a GVF snake for boundary detection. The dark-gray curve is the initial contour, the light-gray curves are iterative runs of the snake algorithm, and the black curve is the final detected boundary.

a few iterations (light-gray curves). Due to the fact that a parametric snake is implemented by using the Lagrangian approach, topological changes, such as splitting of the curve as seen in Figure 6.11, are not allowed during the evolution [175, 117]. Hence, to evolve the GVF snake algorithm for ice floe boundary detection, it is necessary to initialize the contour for each ice floe. Furthermore, because of the explicit representation of the curve, it will be difficult for a parametric snake to segment multiple objects simultaneously [111]. Consequently, the GVF snake algorithm should be carried out on each initial contour one by one.

6.3.1 THE LOCATION OF INITIAL CONTOUR

Even if the GVF snake is faster and less restricted by the initial contour, a proper initial contour is still necessary to ensure the snake evolves to the nearest salient contour. An example is given in Figure 6.12 to illustrate the floe boundary detection results affected by initializing the contour at different locations. In Figure 6.12(a), the initial contour is located in the water, close to the ice boundaries. The snake rapidly detects the boundaries; not the ice but the boundaries of the water region. When initializing the contour at the center of an ice floe, as shown in Figure 6.12(b), the snake accurately finds the boundary after a few iterations, even if the initial contour is some distance away from the floe boundary.

A weak connection will also be detected if the initial contour is located on it, as shown in Figure 6.12(c). However, when the initial contour is located near the floe boundary inside the floe, as shown in Figure 6.12(d), the snake may only find a part of the floe boundary near the initial contour. It should be noted that the curve is always closed, regardless of how it deforms, even in the cases of Figures 6.12(c) and 6.12(d), which appear to result in non-closed curves. This behavior occurs because the area bounded by the closed curve tends to zero.

Figure 6.12 illustrates that, with proper parameters, the snake will find a boundary

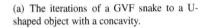

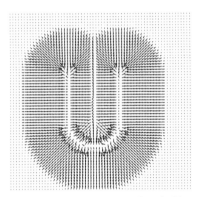

(a) The iterations of a GVF snake to a U-shaped object with a concavity.

(b) GVF force field of the U-shaped object image computed directly from the image gradient.

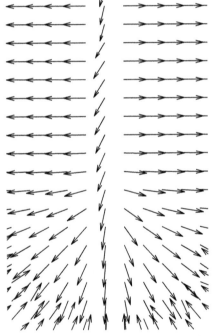

(c) A close look at the GVF force field in the boundary concavity region.

Figure 6.9 The progression of a GVF snake to a boundary concavity. The dark-gray curve is the initial contour, the light-gray curves are iterative runs of the snake algorithm, and the black curve is the final detected boundary.

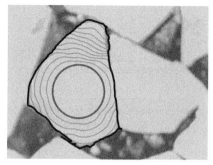

(a) An ice floe image, where the floes are connected to each other.

(b) Evolution of the GVF snake for identifying
the floe boundary.

Figure 6.10   Application of the GVF snake in ice floe boundary detection. The dark-gray curve is the initial contours, the light-gray curves are iterative runs of the GVF snake algorithm, and the black curve is the final detected boundary.

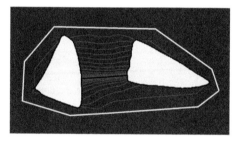

Figure 6.11   Evolution of the GVF snake for multiple ice floes segmentation. The curve cannot split during the evolution. The white curve is the initial contours, the gray curves are iterative runs of the GVF snake algorithm, and the black curve is the final detection result.

regardless of where the initial contour is located. This fact is beneficial for connected floe segmentation. By comparing the results of Figure 6.12, the results where the initial contours are located inside of the floes are most effective, whereas the most efficient case is the one in which the initial contour is in the center of the ice floe. Thus, the initial contour should be located as close to the floe center as possible.

6.3.2   THE SHAPE AND SIZE OF THE INITIAL CONTOUR

In the GVF snake algorithm, the initial contour does not need to be as close to the true boundary as for the traditional snake. However, if the initial contour is too small, it will be slightly "far away" from the floe boundary, and the snake will need more iterations to converge. If the initial contour is further distanced from the floe boundary, the snake may also converge to an incorrect result [174], particularly when the grayscale of the floe is uneven.

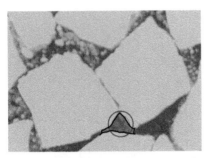

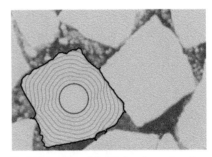

(a) Initial contour 1 located in the water, and the water region boundary is found.

(b) Initial contour 2 located at the center of an ice floe, and the floe boundary is found.

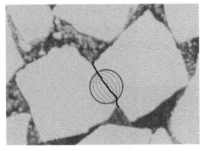

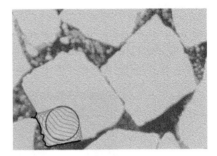

(c) Initial contour 3 located at a weak connection, and the weak connection is found.

(d) Initial contour 4 located near the floe boundary inside the floe, and only a part of the floe boundary is found.

Figure 6.12   Initial contours located at different positions and their corresponding curve evolutions. The dark-gray curves are the initial contours, the light-gray curves are iterative runs of the GVF snake algorithm, and the black curves are the final detected boundaries (Source: Figures from Q. Zhang and R. Skjetne, "Image Processing for Identification of Sea-Ice Floes and the Floe Size Distributions," *IEEE Transactions on Geoscience and Remote Sensing*, 53(5):2913-2924, 2015).

Figure 6.13 serves as an example. Figure 6.13(a) contains some light reflection in the middle of a model sea ice floe [186, 188] where the pixels that belong to the reflection are lighter than the other ice pixels of the floe. Another example is shown in Figure 6.13(d), where the sea ice floe contains some speckle and where the pixels of the speckle are darker. These phenomena will affect the boundary detection when the initial contours (the dark-gray curves in Figures 6.13(b) and 6.13(e)) are too small and not close to the actual boundary. The snake uses many steps (the light-gray curves in Figures 6.13(b) and 6.13(e)) to finally detect a part of the floe boundary (the black curve in Figure 6.13(b)), or to not find the complete boundary because of the speckle (the black curve in Figure 6.13(e)). If we enlarge the initial contours, as shown in Figures 6.13(c) and 6.13(f), the initial contour allows for a faster determination of the correct floe boundary. Therefore, the initial contour should be set as close as possible to the actual floe boundary.

(a) Model sea ice floe image with light reflection.

(b) A small contour initialized at the model sea ice floe center, giving convergence of the snake to the incomplete boundary.

(c) A large contour initialized at the model sea ice floe center, giving convergence of the snake to the correct boundary.

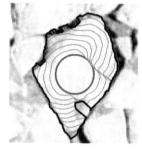

(d) Sea ice floe image with speckle.

(e) A small contour initialized at the sea ice floe center, giving erroneous evolutions of the snake due to the speckle.

(f) A large contour initialized at the sea ice floe center, giving convergence of the snake to the correct boundary.

Figure 6.13 Initial circles with different radii and their curve evolutions. The dark-gray curves are the initial contours, the light-gray curves are iterative runs of the GVF snake algorithm, and the black curves are the final detected boundaries (Source of Figures 6.13(e) and 6.13(f): Figures from Q. Zhang and R. Skjetne, "Image Processing for Identification of Sea-Ice Floes and the Floe Size Distributions," *IEEE Transactions on Geoscience and Remote Sensing*, 53(5):2913-2924, 2015).

6.3.3 AUTOMATIC CONTOUR INITIALIZATION BASED ON THE DISTANCE TRANSFORM

The GVF snake operates on the grayscale image in which the real boundary information, particularly weak boundaries, has been preserved. Moreover, the GVF snake will ensure that the detected boundary is a closed curve. To separate seemingly connected floes into individual ones, the GVF snake algorithm is applied in ice floe boundary detection. However, to start the algorithm, many initial contours are required to identify all individual ice floes. These initial contours should have proper locations, shapes, and sizes. Otherwise, the snake may evolve incorrectly. Therefore, a manual initialization is required in some cases, particularly in crowded floes segmentation. To overcome this problem, an automatic contour initialization algorithm is devised to avoid manual interaction and increase the efficiency of the ice image segmentation method based on the GVF snake algorithm.

As discussed in Section 6.3.1, the initial contours should be located inside the floes to increase the algorithm's effectiveness. In ice image analyses, the ice floes can be separated from water and converted into a binary image by using a thresholding method or k-means clustering method [185, 188]. These methods make it easy to locate the initial contours inside the ice floes. We propose to use the distance transform and its regional maxima to locate the initial contours as close as possible to the floe centers.

As for the definitions of regional and local minima given in Section 5.1, a regional maximum of a grayscale image is defined as a connected component of pixels in a grayscale image with a given value h such that every pixel in its neighborhood has a value strictly lower than h. Similarly, a local maximum is defined as a pixel in a grayscale image if and only if its intensity value is greater or equal to that of any of its neighbors. Every pixel belonging to a regional maximum is a local maximum, but the converse is not true [166]. The set of all regional maxima M_{max} of a grayscale image I can be extracted by using the dilation-based grayscale morphological reconstruction [146]:

$$M_{max} = I - R_I^D(I - 1) \tag{6.57}$$

where $R_I^D(I-1)$ denotes the morphological reconstruction by dilation of I from $I-1$. If the image data type does not support negative values, the following equivalent definition must be considered:

$$M_{max} = I + 1 - R_{I+1}^D(I) \tag{6.58}$$

The distance transform of a binary image, whose elements only have values of '0' and '1' indicating the background and the object pixels, respectively, is the minimum distance from each pixel in the image to the background pixels. Thus, the location of a local maximum belonging to a regional maximum in the distance map ideally corresponds to the center of the object associated with that regional maximum. In many cases, however, multiple regional maxima can be detected for an object, and a regional maximum consists of more than one local maximum (numeral 3 in Figure 6.14(b)). Thus, a dilation operator is used to merge the local maxima within a short distance (within a threshold T_{seed}) of each other. The centers of the dilated regions ('+' in Figure 6.14(b)), which are called 'seeds', are chosen as the locations of the initial contours.

Moreover, to efficiently approach the floe boundary, as discussed in Section 6.3.2, the initial contours should be adapted to the floe sizes. Being unaware of the floe's irregular shape and orientation, the circular shape is chosen as the shape of the initial contour since this shape deforms to the floe boundary more uniformly than other shapes. The radius of the circle is then chosen according to the pixel value at the

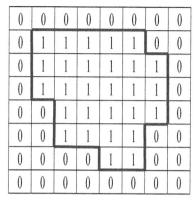

(a) Binary image matrix.

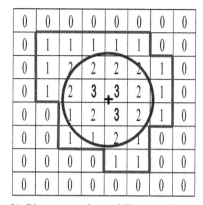

(b) Distance transform of Figure 6.14(a), a regional maximum consisting of three local maxima, a seed, and an initial contour.

Figure 6.14 Contour initialization algorithm based on distance transform (Source: Figures from Q. Zhang and R. Skjetne, "Image Processing for Identification of Sea-Ice Floes and the Floe Size Distributions," *IEEE Transactions on Geoscience and Remote Sensing*, 53(5):2913-2924, 2015).

seed in the distance map. This ensures that the initial circle is contained strictly inside the floe, as shown as the circle in Figure 6.14(b)[4].

After initializing the contours, the GVF snake algorithm is run on each contour to identify the floe boundary. Superimposing all the boundaries onto the binarized ice image results in the separation of the connected ice floes. The final segmentation result of Figure 6.15(e) is shown in Figure 6.15(f). Note that the edge pixels are specifically labeled as "residue ice" for special handling in subsequent use.

6.4 ICE IMAGE SEGMENTATION

According to Section 6.3, the following Algorithm 1 is proposed to segment the ice image. First, the GVF is derived from the grayscale input image. Then, the ice pixels are separated from water pixels, and the image is converted into a binary one. Next, the distance transform is applied to the binary image, and the seeds and radii are found. Finally, based on the seeds and radii, the circles are initialized and the snake algorithm is run. The pseudocode of the proposed algorithm is given in Algorithm 1, and its procedure carried out on a sea ice floe image is shown in Figure 6.15.

It should be noted that the regional maxima whose distance is larger than the given threshold T_{seed} will not be merged into one (dilated) region. This means that some

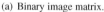

[4]Note: since the city-block distance is used here due to its ability of generating a moderate number of regional maxima between the Euclidean and chessboard distance transforms as discussed in Section 5.1.2, the pixel values at seeds must be divided by $\sqrt{2}$ to obtain the circle radii.

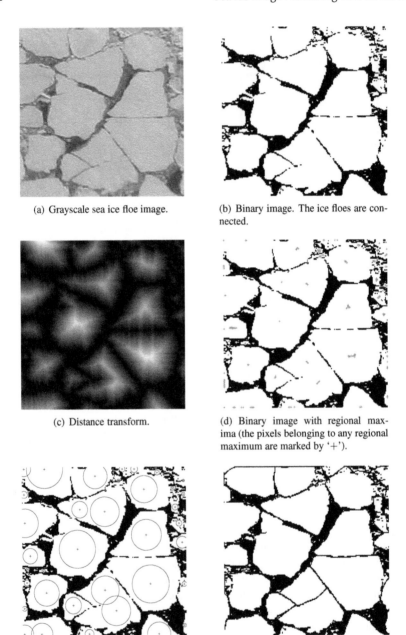

(a) Grayscale sea ice floe image.

(b) Binary image. The ice floes are connected.

(c) Distance transform.

(d) Binary image with regional maxima (the pixels belonging to any regional
maximum are marked by '+').

(e) Binary image with seeds (marked by
'+') and initial contours (circles).

(f) Segmentation result. The connected
ice floes are separated.

Figure 6.15 The procedure of ice segmentation algorithm (Source of Figures 6.15(a),
6.15(e), and 6.15(f): Figures from Q. Zhang and R. Skjetne, "Image Processing for Identification of Sea-Ice Floes and the Floe Size Distributions," *IEEE Transactions on Geoscience
and Remote Sensing*, 53(5):2913-2924, 2015).

Algorithm 1 Ice image segmentation

Input: Ice image

Start algorithm:

1: $GVF \leftarrow$ GVF derived from grayscale of input image
2: $SEGMENTATION \leftarrow$ binary ice image
3: $D \leftarrow$ distance map of $SEGMENTATION$
4: $M \leftarrow$ regional maximum of D
5: $S \leftarrow$ Seeds of $SEGMENTATION$ found by merging the regional maxima in M within a short distance (T_{seed})
6: **for** each seed $s \in S$ **do**
7:     $r \leftarrow$ pixel value at s in D
8:     $c \leftarrow$ initial contour locate at s with its radius r
9:     $B \leftarrow$ boundary detected by performing the GVF snake algorithm on c
10:     $SEGMENTATION \leftarrow SEGMENTATION$ with B superimposed
11: **end for**
12: **return** $SEGMENTATION$

Output: Segmented ice image

---

floes may have more than one seed. However, two or more seeds for one ice floe will not affect its boundary detection, but it may increase the computational time.

6.5   DISCUSSION

6.5.1   STOPPING CRITERION FOR THE SNAKE

A stopping criterion for the iterative process of the snake was not given by either the traditional snake [75] or the GVF snake algorithm [174]. Without a stopping criterion, the snake will keep evolving even though it has converged to the desired boundaries. A simple solution to this issue is made by limiting the number of iterations. That is, the snake should stop its evolution if the maximum number of iterations has been exceeded. Higher values of the maximum iteration might result in many unnecessary iterations, using unnecessary computational time, whereas lower values might impede the snake's ability to converge to its minimum (the desired image feature). Hence, how to choose an appropriate value for the upper limit on the number of iterations of the snake's evolution is a challenge for this approach.

Another choice for stopping the snake's evolution can be done by checking if the total snake energy functional value for each new iteration changes sufficiently. If the reduction in energy from one step to the next is below a small threshold, then one can assume that a minimum has been reached [124]. Note though, that in some cases the evolution of the snake may be very small during some intermediate steps before suddenly growing large again. As seen in Figure 6.9(a), for instance, the snake evolves very slowly in the channel region of the U-shaped object, but after that it evolves much faster into the boundary concavity. So the issue then becomes the selection of this small threshold.

6.5.2   GVF CAPTURE RANGE

In addition to the regularization parameter μ, the GVF snake has another parameter: the GVF field iteration, as compared to the traditional snake. The GVF field iteration is the internal loop counter used to calculate the GVF force field of the image. This controls the capture range of the force field. The more iterations needed for calculating the GVF field, the larger capture the range becomes for the force field. However, this comes at the expense of more computation time. As seen in Figure 6.16, where the GVF force fields on a large and a small circle are calculated via a different number of iterations, the capture range of the force field is significantly enlarged when the GVF field iteration increases. As a result, the initial contour can be placed a bit farther away from the object boundary, and the snake will converge faster to the boundary when increasing the GVF field iteration, as seen in Figure 6.17.

Usually a large object (or image) size requires a large GVF iteration number. Figure 6.16(b) shows that 5 iterations are sufficient for the 9-pixel wide diameter circle to be "filled up" with the GVF capture range. Then the initial contour can evolve under the influence of external forces no matter where it is initialized inside the small circle. On the other hand, Figure 6.16(c) implies that 30 iterations are still too small to generate enough GVF capture range to cover the inside of the larger circle, whose diameter is 61-pixel wide. Thus, it should avoid initializing a small contour near the center of the large circle, outside the capture range. Otherwise, no external forces will act on the snake contour. Hence, the GVF iteration number is typically set proportional to the size of the object (image) to reduce the requirements in initial contours [61].

When the GVF force fields are calculated via the same number of iterations, the external forces near weak edges are weaker than those near strong edges. If the number of iterations for deriving the GVF field is too large, due to the inherent competition of the diffusion process, the capture range of the strong edges may dominate the external force field, and the external forces near the weak edges, which are close to the stronger ones, will be too weak to pull the snake toward the desired weak boundaries. As a result, the snake is likely to pass over the weak edge and terminate at the corresponding strong edge [84].

As seen in Figure 6.18, for instance, the GVF snake-based segmentation algorithm can separate the model sea ice floes correctly under appropriate GVF capture range. When enlarging the capture range, the external force vectors near the weak edge are overwhelmed by the vectors near the strong edge as a result of the diffusion, and no force is then present to pull the snake to the weak edge, with the result that the weaker edge is lost and the small model sea ice floe is merged into the big one. In contrast, taking Figure 6.19, for example, under the smaller GVF capture range, the external force vectors of the floe boundary are not strong enough to overwhelm the vectors of the noise. So the external forces generated from the noise can block the snake from propagating to the floe boundary, and the floe is thereby divided into several small pieces.

Usually, weak edges tend to be more difficult to detect when increasing the GVF capture range, which results in under-segmentation. If, on the other hand,

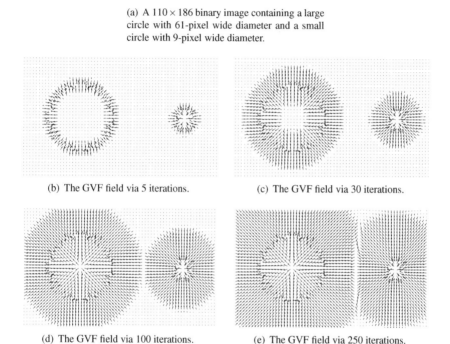

(a) A 110×186 binary image containing a large circle with 61-pixel wide diameter and a small circle with 9-pixel wide diameter.

(b) The GVF field via 5 iterations.

(c) The GVF field via 30 iterations.

(d) The GVF field via 100 iterations.

(e) The GVF field via 250 iterations.

Figure 6.16   The GVF force fields via different iterations.

the capture range is decreased, the noise may be enhanced, which may induce over-segmentation.

Note that a small object may be mistaken as noise if it is located close to a relatively large object. Then the snake may fail to find the boundary of the small object when the GVF capture range is large.

6.5.3   BORDER EFFECTS

According to Section 6.5.2, the GVF capture range derived by a uniform parameter for the whole image sometimes cannot represent an overall ice image well. Thus, processing the overall ice image locally is typically recommended in order to identify all of the boundaries. A common way is to divide the overall image into several sub-

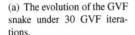

(a) The evolution of the GVF snake under 30 GVF iterations.

(b) The evolution of the GVF snake under 250 GVF iterations.

Figure 6.17 The evolutions of the GVF snake to the boundary of the large circle shown in Figure 6.16(a) under different GVF iterations.

images and apply the GVF snake-based segmentation algorithm to each sub-image. However, attention should be paid when dividing the overall image, due to the fact that the segmentation of the floes situated at the border of the analyzed image is more prone to error. This phenomenon occurs because these floes are usually incompletely "cut" by the image border.

On the one hand, the image border might influence the evolution of the snake. An example is shown in Figure 6.20. When the floe is complete, the snake propagates to the floe boundary and determines the entire boundary. If the floe is incomplete, the convergence of the snake might be impeded when meeting an image border, because the snake cannot split and the snake will then stop at the image border. The result is that the GVF snake-based segmentation algorithm may over-segment the ice floe, which connects to the border of the image.

On the other hand, the incompleteness of the floe might influence the GVF force field of the floe and then affect the final segmentation result. As shown in Figure 6.21(b), the GVF force field of a complete floe radiates from the floe center to its boundary. The segmentation algorithm then successfully finds all the boundary pixels shown in Figure 6.21(c). Contrary to this, when part of the floe boundary is lost due to an image border, the GVF force field of this incomplete floe rather radiates from the image border to the remaining floe boundary, as shown in Figure 6.21(e). Then all the points on the snake will be pushed toward the remaining boundary along the direction of the force vectors. According to the GVF snake-based segmentation algorithm, the initialization of the snake contour for the cut floe is based on its incomplete shape. Hence, it might be unable to initialize the contour close to the floe edges that connect to the image border. If these edges are very weak, it tends to be much easier for the snake to pass over them, with the result that the segmentation by the snake algorithm fails.

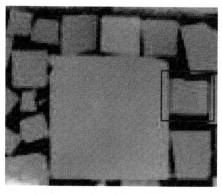

(a) A grayscale model sea ice image.

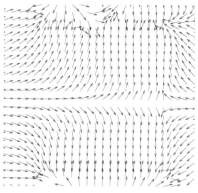

(b) The GVF field of the block area in Figure 6.18(a) via 800 iterations. The external force vectors near the weak edge are overwhelmed by the vectors near the strong edge.

(c) Under-segmentation result of Figure 6.18(a) via large GVF capture range (800 iterations for GVF).

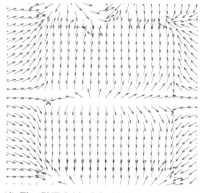

(d) The GVF field of the block area in Figure 6.18(a) via 80 iterations.

(e) Correct segmentation result of Figure 6.18(a) via small GVF capture range (80 iterations for GVF).

Figure 6.18   Under-segmentation due to the large GVF capture range.

(a) A grayscale model sea ice image.

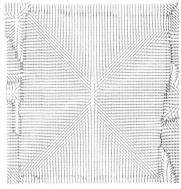

(b) The GVF field of Figure 6.19(a) via 600 iterations.

(c) Correct segmentation result of Figure 6.19(a) via large GVF capture range (600 iterations for GVF).

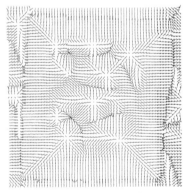

(d) The GVF field of Figure 6.19(a) via 60 iterations. The noise is enhanced.

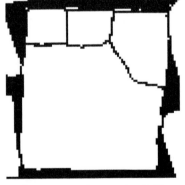

(e) Over-segmentation result of Figure 6.19(a) via small GVF capture range (60 iterations for GVF).

Figure 6.19 Over-segmentation due to the small GVF capture range.

(a) Model sea ice image contained a complete floe.

(b) The propagation of the snake for the complete floe. The entire floe boundary is found.

(c) Segmentation result of Figure 6.20(a).

(d) The model sea ice floe in Figure 6.20(a) is being "cut" by the image border.

(e) The propagation of the snake for the incomplete floe. The snake is impeded by the image border.

(f) Segmentation result of Figure 6.20(d).

Figure 6.20 The influence of image border on the snake's evolution.

 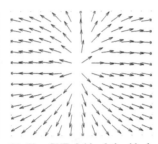 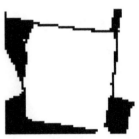

(a) Model sea ice image con-   (b) The GVF field of the block   (c) Segmentation result of Fig-
tained a complete floe.   region in Figure 6.21(a) radiates   ure 6.21(a). All the boundary
                                 from the floe center to its bound-   pixels are found.
                                 aries.

 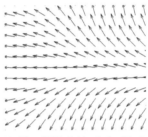 

(d) The model sea ice   (e) The GVF field of the block   (f) Segmentation re-
floe in Figure 6.21(a)   region in Figure 6.21(d) radiates   sult of Figure 6.21(d).
is being "cut" by an   from the image border to the floe   A part of the floe
image border.   boundaries.   boundary is lost.

Figure 6.21   The influence of image border on the GVF force field (Source of Figure 6.21(c)
and 6.21(f): Figures from Q. Zhang, R. Skjetne, I. Metrikin, and S. Løset, "Image Processing
for Ice Floe Analyses in Broken-ice Model Testing," *Cold Regions Science and Technology,*
111:27-38, 2015).

7 Sea Ice Type Identification

A GVF snake-based algorithm was introduced in Chapter 6 to separate seemingly connected floes into individual ones. In this chapter, an ice shape enhancement algorithm is employed. This can be used after the segmentation of the floes to preserve the ice floe shapes and accomplish the identification of individual ice floes and brash ice pieces. Once the ice floes and brash ice are identified, different types of sea ice (i.e., ice floe, brash ice, and slush) can be classified, and the floe size distribution can be calculated from the resulting data.

7.1 ICE SHAPE ENHANCEMENT

After segmentation by using the GVF snake algorithm, some segmented ice floes may contain holes or smaller ice pieces within them due to noise and speckle, as shown in Figure 7.1, and the shapes of the detected floes are rough. Thus, the morphological cleaning [145] is used after the segmentation to smoothen the shapes of the ice floes. Additionally, the technique of connected component labeling, which is used to extract each individual segmented ice floe, and the technique of hole filling are also required for ice shape enhancement.

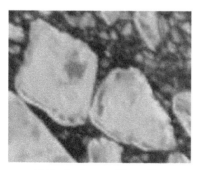 

(a) Ice floe image with speckle. (b) Segmentation result of Figure 7.1(a).

Figure 7.1 The segmentation of an ice floe image with speckle. The segmented floes contain holes and smaller ice pieces within their shapes (Source: Figures from Q. Zhang and R. Skjetne, "Image Techniques for Identifying Sea-Ice Parameters," *Modeling, Identification and Control*, 35(4):293-301, 2014).

Note that since the output of the GVF snake segmentation method (Algorithm 1) is the segmented binary image, the ice shape enhancement algorithm will be a binary operation.

7.1.1 MORPHOLOGICAL CLEANING

Morphological cleaning is a combination of first morphological closing and then morphological opening on an image [145]. Both binary closing and opening operations can smooth the contours of objects, yielding results that are similar to the original shapes of the objects but with different level of details. The closing operation is able to close narrow cracks, fill long thin channels, and eliminate the holes that are smaller than the structuring element. The opening operation is able to break thin connections between objects, remove small protrusions, and eliminate complete regions of an object that cannot contain the structuring element. Figure 7.2 shows the effects of binary morphological closing and opening, illustrating the morphological cleaning.

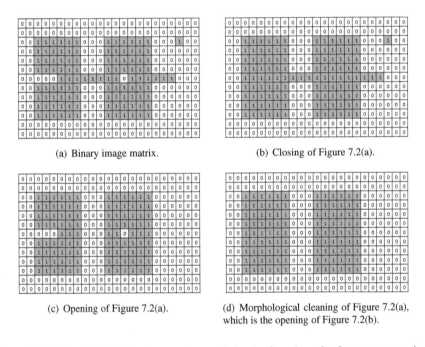

(a) Binary image matrix.   (b) Closing of Figure 7.2(a).

(c) Opening of Figure 7.2(a).   (d) Morphological cleaning of Figure 7.2(a),
                                           which is the opening of Figure 7.2(b).

Figure 7.2   Morphological closing, opening, and cleaning by using a 2×2 square structuring element.

7.1.2 CONNECTED COMPONENT EXTRACTION AND LABELING

A connected component is a set of pixels in an image that are all connected to each other. The extraction and labeling of connected components, such as objects, from a binary image plays a central role to many automated image processing applications. Morphological operations can be used to extract and label connected components in a binary image.

Let A denote a binary image set containing one or more connected components, in which all object pixels are 1-valued and background pixels are 0-valued. To extract a connected component, the connected component labeling algorithm starts with an arbitrary pixel p in each component in A. Then the algorithm grows it by performing the following iterative dilation procedure [49]:

$$X_k = (X_{k-1} \oplus B) \cap A \quad k = 1, 2, 3, \cdots \tag{7.1}$$

where B is a suitable structuring element, and the initial step $X_0 = \{p\}$ is an array of the same size as A with 0-valued elements except at p, where it is 1-valued. The iteration procedure terminates at step k when meeting the condition that $X_k = X_{k-1}$, at which point X_k contains all the connected components.

In order to distinguish different connected components in an image, distinctive labels must be assigned to each extracted connected component. Thus, an arbitrary object pixel that is not yet assigned to any connected component can be picked up as the starting pixel p, and then a unique label is assigned to the resulting connected component that is extracted [12].

Figure 7.3 gives a simple example of the connected component labeling algorithm, where the structuring element B is a square matrix of size 3×3, as shown in Figure 7.3(a). In this case, the connected component uses 8-connectivity, where the adjacent pixels of an object have at least one common corner, and the algorithm completes at the 8th iteration. When using a 4-connected component, in which the adjacent pixels of an object are required to have a common side, the cross structuring element shown in Figure 7.4(a) is chosen as the structuring element B. Then, the connected component labeling algorithm gives a different result, as shown in Figure 7.4, based on the same binary image matrix and initial step as in Figure 7.3.

7.1.3 HOLE FILLING

A hole is a background region surrounded by the object's boundary pixels. Let A denote a binary image set containing 1-valued object boundary pixels, where each object boundary encloses one or more 0-valued holes. Then the objective of the hole filling algorithm is to set value 1 to all the hole pixels contained within the object boundaries in A.

7.1.3.1 Iterative dilation-based hole filling algorithm

Similar to the connected component labeling algorithm, the hole filling algorithm can start with an arbitrary pixel p in each hole in A and use an iterative dilation procedure. This will grow the region of pixel p and then fill the holes within the object boundary [49], according to:

$$X_k = (X_{k-1} \oplus B) \cap A^c \quad k = 1, 2, 3, \cdots \tag{7.2}$$

where B is the cross structuring element, and $X_0 = \{p\}$ is the initial step. The algorithm terminates at step k if $X_k = X_{k-1}$, at which point X_k contains all the holes that

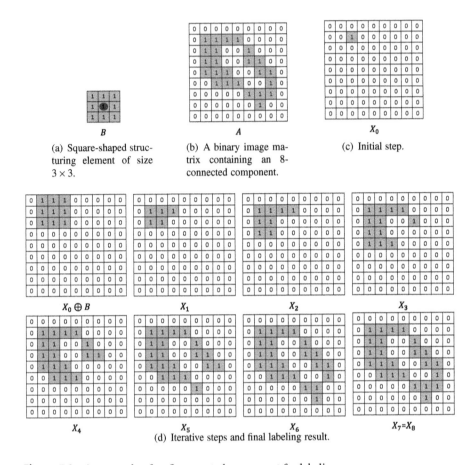

Figure 7.3   An example of an 8-connected component for labeling.

need to be filled. The union of X_k and A yields the final result of filled holes and their boundaries. A simple example of the hole filling algorithm is shown in Figure 7.5. In this case, the algorithm finishes at the 7th iteration and gives the final filling result as the union of the original binary image A with the iteration result X_7.

As seen in Figure 7.6(b), the object contains two 4-connected holes. With an initial hole pixel p, only one 4-connected hole is determined and filled when using the cross structure element. In this case, the 3×3 square-shaped structure element cannot help to improve this, because these two 4-connected holes, which can be treated as one 8-connected hole, are 8-connected to the background region. All 0-valued background pixels are then treated as one 8-connected "hole" by this algorithm, as seen in Figure 7.7. It follows that in order to fill all holes in an image, this algorithm requires knowing whether the initial 0-valued pixel p belongs to a hole within an object or to the background. This makes it difficult to automatically fulfill this task.

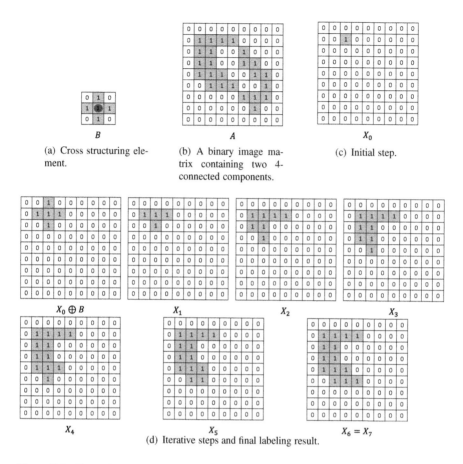

(a) Cross structuring element.

(b) A binary image matrix containing two 4-connected components.

(c) Initial step.

(d) Iterative steps and final labeling result.

Figure 7.4 Example of 4-connected component labeling of the same binary matrix and initial step as for Figure 7.3(b).

A fully automatic approach is thus aimed for to increase the level of autonomy in the overall algorithm.

7.1.3.2 Morphological reconstruction-based hole filling algorithm

The dilation-based binary morphological reconstruction extracts the connected components of the mask image "marked" by the marker image that is contained in the mask image [166]. It can thus be used to develop a fully automatic hole filling algorithm [48].

Let F be a binary image containing one or more objects with holes inside. Let F_m be a marker image that is 0 everywhere except on the image domain border, where it

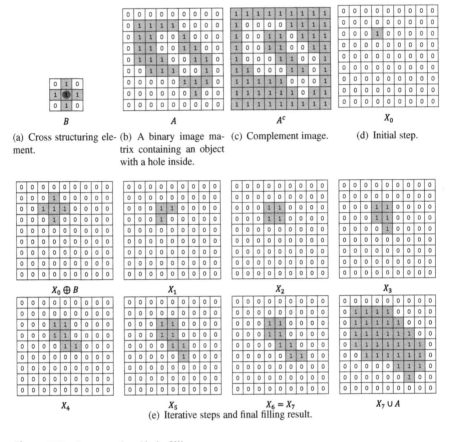

(a) Cross structuring ele- (b) A binary image ma- (c) Complement image.   (d) Initial step.
ment.   trix containing an object
                           with a hole inside.

(e) Iterative steps and final filling result.

Figure 7.5 An example of hole filling.

is set to $1 - F$:

$$F_m(x,y) = \begin{cases} 1 - F(x,y) & \text{if } (x,y) \text{ is on the border of } F \\ 0 & \text{otherwise} \end{cases} \tag{7.3}$$

Then

$$H = \left[R_{F^c}^D (F_m) \right]^c \tag{7.4}$$

has the effect of filling the holes in F. This fully automatic hole filling algorithm is based on the morphological reconstruction by dilation. Its process is illustrated in Figure 7.8.

7.1.4   ICE SHAPE ENHANCEMENT ALGORITHM

Using the morphological cleaning, all the segmented ice pieces derived from the ice floe segmentation result of Algorithm 1 are first labeled and arranged from small to

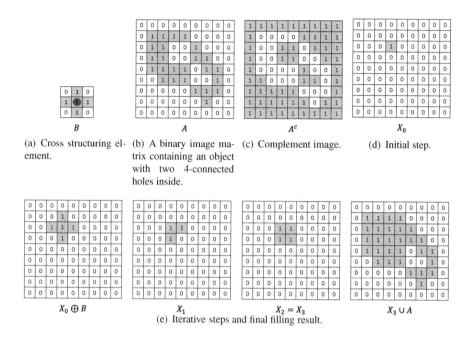

(a) Cross structuring element.

(b) A binary image matrix containing an object with two 4-connected holes inside.

(c) Complement image.

(d) Initial step.

(e) Iterative steps and final filling result.

Figure 7.6   An example of unsuccessful 4-connected hole filling by using the cross structuring element.

large. Then the morphological cleaning with a proper structuring element and hole filling algorithms are performed to the arranged ice pieces in sequence. The pseudocode of the proposed ice shape enhancement algorithm is concluded in Algorithm 2. The ice shape enhancement result of Figure 7.1(b) is shown in Figure 7.9.

It should be noted that the arrangement of ice pieces in order of increasing size is required for the morphological cleaning and hole filling algorithms. Otherwise, the smaller ice piece contained in a larger ice floe may not be removed.

7.2   GENERAL SEA ICE IMAGE PROCESSING

Sea ice typically has a wide variability of ice floe sizes, together with content of brash ice and slush and, possibly, a snow cover. Various types of ice and irregular floe sizes and shapes are challenges to sea ice image processing. Section 6.4 proposes a method based on the GVF snake algorithm to separate seemingly connected ice floes, and Section 7.1 adopts the morphological cleaning to enhance the shapes of the segmented ice. The main focus of this section is on identifying the non-ridged ice floe in the marginal ice zone based on those methods. A general view of processing a sea ice image to classify different sea ice types and obtain the floe size distribution, is given in this section.

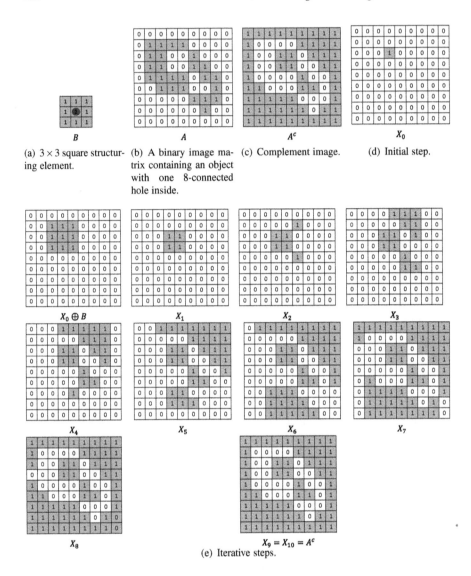

(a) 3×3 square structuring element.

(b) A binary image matrix containing an object with one 8-connected hole inside.

(c) Complement image.

(d) Initial step.

(e) Iterative steps.

Figure 7.7 An example of unsuccessful hole filling of the same binary matrix as Figure 7.6(b) by using a 3×3 square structuring element.

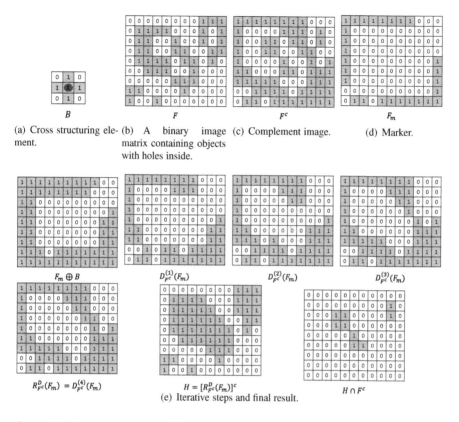

(a) Cross structuring element.

(b) A binary image matrix containing objects with holes inside.

(c) Complement image.

(d) Marker.

(e) Iterative steps and final result.

Figure 7.8 An example of hole filling by using the morphological reconstruction.

Algorithm 2 Ice shape enhancement

Input: Ice segmentation *SEGMENTATION* from Algorithm 1

Start algorithm:

1: *PIECES* ← labeled regions in *SEGMENTATION* arranged from small to large
2: *BW* ← empty black image
3: **for** each labeled region *piece* ∈ *PIECES* (from small to large) **do**
4:     *piece* ← morphological clean and fill hole
5:     *BW* ← *BW* with *piece* superimposed and labeled
6: **end for**
7: *IDENTIFICATION* ← labeled regions in *BW*
8: **return** *IDENTIFICATION*

Output: Individual ice piece identification

Figure 7.9   Ice shape enhancement result of Figure 7.1(b). The holes and smaller ice pieces inside larger floes are removed (Source: Figure from Q. Zhang and R. Skjetne, "Image Techniques for Identifying Sea-Ice Parameters," *Modeling, Identification and Control*, 35(4):293-301, 2014).

7.2.1   SEA ICE PIXEL EXTRACTION

Because sea ice is whiter than water, the pixel values differ under normal conditions. Ice pixels have higher intensity values than those belonging to water in a uniformly illuminated ice image. Therefore, ice pixels can be extracted by using some thresholding method. As shown in Figure 7.10(b), most of the ice in Figure 7.10(a) is identified by the Otsu thresholding method. However, only the "light ice" pixels with larger pixel intensity values than the threshold are identified. The "dark ice" pixels with intensity values between the threshold and water, such as for ice pieces under the water surface, may not be identified. They are considered to be water according to the thresholding method. Since both "light ice" and "dark ice" pixels are required for an accurate analysis, we aim to distinguish "dark ice" from open water. For this, the k-means clustering algorithm can be applied.

The image is divided into three or more clusters using the k-means algorithm. The cluster with the lowest average intensity value is considered to be water, while the other clusters are considered to be ice, as shown in Figure 7.10(c). The "dark ice" is obtained by comparing the difference between Figure 7.10(b) and 7.10(c), as shown in Figure 7.10(d). We will see later that creating the individual "light ice" and "dark ice" layers is advantageous for the computation of the initial contours for the GVF snake algorithm.

Note that the results of the ice pixel detection by using the same method but with different multilevel may be similar to each other. As an example, consider Figure 3.11(c) and Figure 3.12(b), which show ice pixel detection results by using the k-means clustering with two and three clusters, respectively. The difference between these two resulting images is too minor to be able to determine the "dark ice" pixels. Therefore, it is necessary to use two different methods for the extractions of the sea ice pixels in order to widen the gap between the results.

Furthermore, if we use the k-means clustering method to detect the "light ice" pixels, as seen in Figure 3.11(c), then it becomes difficult to initialize the contours for the GVF snake algorithm, in order to separate the connected ice floes in the area where the sea ice are crowded. Comparing Figures 3.11(b) and 3.11(c), we see that the Otsu thresholding method detects less pixels than the k-means clustering method. However, the under-detected ice pixels by the Otsu method results in more "holes" in the binarized image, and this is essential for the further initialization of the contours for the GVF snake algorithm. Those under-detected ice pixels can next be compensated by the detected "dark ice" pixels when using the additional k-means clustering method with three clusters. Note that using the Otsu multithresholding method to detect "dark ice" pixels will require more computation time. Therefore, we have proposed to use the Otsu thresholding method to detect the "light ice" pixels, then use the k-means clustering method with three or more clusters to find the "dark ice" pixels, and thereafter use the "light ice" and "dark ice" layers individually to calculate the initialization of the contours for the GVF snake algorithm. Details of this is described next.

7.2.2 SEA ICE EDGE DETECTION

The sea ice pixel extraction results in a "light ice" image, as seen in Figure 7.10(b), and a "dark ice" image, as seen in Figure 7.10(d)). In order to start the GVF snake algorithm for sea ice image processing, the contours are initialized in both the "light ice" image layer and "dark ice" image layer to obtain a more accurate result and reduce the computational time. Then the GVF snake algorithm is run to individually derive "light ice" segmentation as seen by the white ice pieces in Figure 7.11, and the "dark ice" segmentation, as seen by the gray ice pieces in Figure 7.11. Collecting all the ice pieces in the "light ice" and "dark ice" segmented image layers results in the final segmented image, as exemplified in Figure 7.11. It should be noted that the "light ice" and the "dark ice" should be labeled differently in the final segmented image (for example, setting "light ice" pixels value 1, "dark ice" pixels value 0.5, and water pixels value 0). Otherwise, it may be impossible to separate some "light ice" and "dark ice" pieces if they are connected.

The pseudocode of the proposed sea ice edge detection algorithm is concluded in Algorithm 3.

7.2.3 SEA ICE SHAPE ENHANCEMENT

Figure 7.12(a), which is extracted from Figure 7.10(a), shows an ice floe with speckle. Because of the uneven grayscale of the ice floe, the lighter part of the floe is considered as "light ice" (the white pixels in Figures 7.12(b) and 7.12(c)), while the darker part is considered as "dark ice" (the gray pixels in Figures 7.12(b) and 7.12(c)). This means the ice floe, as shown in Figure 7.12(b), cannot be completely identified when it has both "light ice" pixels and "dark ice" pixels. Therefore, the ice shape enhancement is particularly important for the sea ice image processing.

If we perform the ice shape enhancement to the "light ice" segmentation and "dark ice" segmentation independently, there will be overlap between the resulting individual light ice piece identification and individual dark ice piece identification.

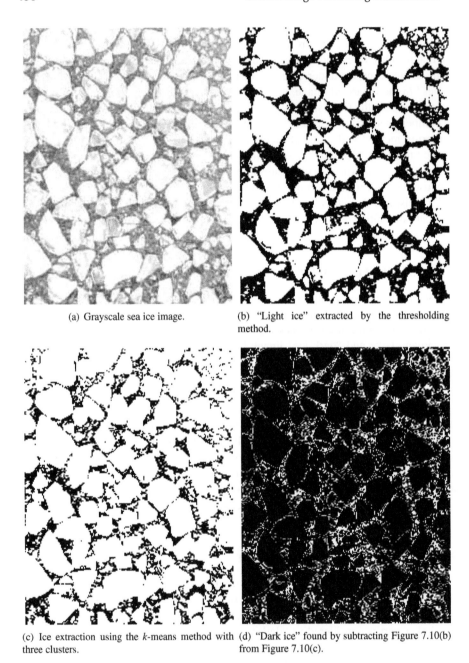

(a) Grayscale sea ice image.

(b) "Light ice" extracted by the thresholding method.

(c) Ice extraction using the *k*-means method with three clusters.

(d) "Dark ice" found by subtracting Figure 7.10(b) from Figure 7.10(c).

Figure 7.10   Sea ice pixel extraction (Source: Figures from Q. Zhang and R. Skjetne, "Image Processing for Identification of Sea-Ice Floes and the Floe Size Distributions," *IEEE Transactions on Geoscience and Remote Sensing*, 53(5):2913-2924, 2015).

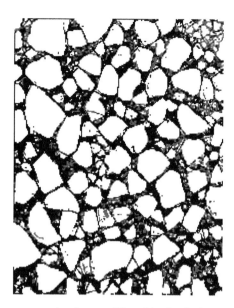

Figure 7.11 Sea ice segmentation image. The white ice is the segmentation result from the "light ice" in Figure 7.10(b), and the gray ice is the segmentation result from the "dark ice" in Figure 7.10(d) (Source: Figure from Q. Zhang and R. Skjetne, "Image Processing for Identification of Sea-Ice Floes and the Floe Size Distributions," *IEEE Transactions on Geoscience and Remote Sensing*, 53(5):2913-2924, 2015).

Algorithm 3 Sea ice edge detection

Input: Sea ice image

Start algorithm:

1: $GVF \leftarrow$ GVF derived from grayscale of input image

2: $ICE \leftarrow$ binary ice image by the k-means clustering method

3: $LIGHT \leftarrow$ binary "light" ice image by the thresholding method

4: $DARK \leftarrow ICE - LIGHT$

5: $SEG_L \leftarrow$ ice floe segmentation (Algorithm 1) on $LIGHT$

6: $SEG_D \leftarrow$ ice floe segmentation (Algorithm 1) on $DARK$

7: $SEGMENTATION_{sea\ ice} \leftarrow SEG_L + SEG_D$ (SEG_L and SEG_D are labeled differently in $SEGMENTATION_{sea\ ice}$)

8: **return** $SEGMENTATION_{sea\ ice}$

Output: Sea ice segmentation

This means that some ice pixels may be identified as belonging to different ice floes, with the risk that large ice floes are still incomplete.

To perform the ice shape enhancement algorithm to the sea ice segmentation image, all the detected ice pieces, including both segmented "light ice" and "dark ice" pieces, should be labeled as one input to the step of ice shape enhancement. A disk-shaped structuring element is chosen for the ice shape enhancement. The radius of this disk can be automatically adapted to the size of each ice piece according to some rule, such as Equation 7.5:

$$r = \begin{cases} r_1 & \text{if } size_{ice} < size_{th} \\ r_2 & \text{if } size_{ice} \geq size_{th}, \end{cases} \tag{7.5}$$

where r is the radius of the disk. It is equal to r_1 when the size of the ice piece $size_{ice}$ is less than a threshold $size_{th}$. Otherwise, it is set equal to r_2. This ensures that the shapes of each ice piece can be better preserved according to their sizes.

The pseudocode of the sea ice shape enhancement algorithm is concluded in Algorithm 4. This process will ensure the completeness of each ice floe and that smaller ice floes or brash pieces contained within a larger ice floe are merged with the larger floe, as shown in the shape enhancement result in Figure 7.12(c).

Algorithm 4 Sea ice shape enhancement

Input: Sea ice segmentation $SEGMENTATION_{sea\ ice}$ from Algorithm 3
Start algorithm:
  1: $PIECES_{sea\ ice} \leftarrow$ labeled regions in $SEGMENTATION_{sea\ ice}$
  2: $IDENTIFICATION_{sea\ ice} \leftarrow$ ice shape enhancement (Algorithm 2)
  3: **return** $IDENTIFICATION_{sea\ ice}$
Output: Individual sea ice piece identification

(a) Ice floe image extracted from Figure 7.10(a).
(b) Segmentation result of Figure 7.12(a).
(c) Shape enhancement result of 7.12(b).

Figure 7.12 Sea ice shape enhancement. The white pixels are the "light ice" pixels, and the gray pixels are the "dark ice" pixels (Source: Figures from Q. Zhang and R. Skjetne, "Image Processing for Identification of Sea-Ice Floes and the Floe Size Distributions," *IEEE Transactions on Geoscience and Remote Sensing*, 53(5):2913-2924, 2015).

After shape enhancement, collecting and labeling all the cleaned ice pieces in different colors based on their areas (e.g., calculated by counting the pixel number of the floe) according to

$$\mathbf{Color}(p) = \begin{cases} 0 & \text{if } p \notin ICE_{SEA} \\ C_1(1 - \exp(-area(i)/C_2)) & \text{if } p \in ice_{sea}(i) \end{cases} \qquad (7.6)$$

where $ICE_{SEA} = \{ice_{sea}(1), ice_{sea}(2), ice_{sea}(3), \cdots\}$ is a set of identified (cleaned) sea ice pieces derived from Algorithm 4, $ice_{sea}(i) \in ICE_{SEA}$ is ice piece i, and $area(i)$ is its area. The constants are proposed set to $C_1 = 10000$ and $C_2 = 1000$. This labeling results in Figure 7.13, where smaller ice pieces are blue and larger ice pieces are red. The ice piece positions, found by averaging the positions of the pixels of each ice piece, are denoted by the black dots.

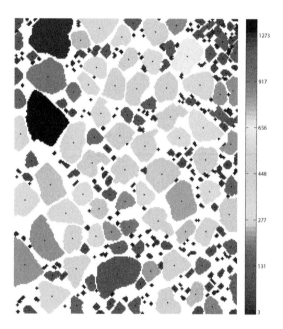

Figure 7.13   Labeled and colorized ice pieces (Source: Figure from Q. Zhang and R. Skjetne, "Image Processing for Identification of Sea-Ice Floes and the Floe Size Distributions," *IEEE Transactions on Geoscience and Remote Sensing*, 53(5):2913-2924, 2015).

7.2.4   SEA ICE TYPES CLASSIFICATION AND FLOE SIZE DISTRIBUTION

In ice engineering, one generally assumes that brash ice has a dampening effect in models for calculating ice pressure and ice forces. It is therefore important to estimate the distribution of ice floes and content of brash ice. According to [113], brash ice is considered as floating ice fragments no more than 2 m across. To distinguish brash ice from ice floes in our algorithm, we define a brash-ice threshold parameter (pixel number, area, or characteristic length) that can be tuned for each application.

The ice pieces with sizes larger than the threshold are considered to be ice floes, while smaller pieces are considered to be brash ice. The remaining ice pixels are labeled as slush. The sea ice is then divided into three classes: ice floes, brash ice, and slush. Therefore, the sea ice classification results in four layers of a sea ice image (using Figure 7.10(a) as an example): ice floe (Figure 7.14(a)), brash ice (Figure 7.14(b)), slush (Figure 7.14(c)), and water (Figure 7.14(d)). Based on the four layers, a total of 154 ice floes and 189 brash ice pieces are identified from Figure 7.10(a). The coverage percentages are 60.52% ice floe, 3.34% brash ice, 16.03% slush, and 20.11% water. The histogram of ice floe size distribution grouped by pixel numbers is shown in Figure 7.15.

The residue ice, which is the detected edge pixels between the connected floes, was in this example considered as slush (since there often is an edge layer of slush ice between two ice floes) and included in Figure 7.14(c). However, the residue ice, as shown in Figure 7.16, can also simply be identified as "residue ice" and defined specifically by the user and the application of the data.

The pseudocode of the ice types classification algorithm is concluded in Algorithm 5.

Algorithm 5 Sea ice types classification algorithm

Input: Individual sea ice piece identification $IDENTIFICATION_{sea\ ice}$ from Algorithm 4, Threshold T_{floe}, Binary ice image ICE by the k-means clustering method from Algorithm 3

Start algorithm:
1: $FLOE \leftarrow$ labeled regions in $IDENTIFICATION_{sea\ ice}$ with the sizes equal to or larger than T_{floe}
2: $BRASH \leftarrow$ labeled regions in $IDENTIFICATION_{sea\ ice}$ with the sizes smaller than T_{floe}
3: $PIXEL_{sea\ ice} \leftarrow IDENTIFICATION_{sea\ ice} \cup ICE$
4: $SLUSH \leftarrow PIXEL_{sea\ ice} \cap (IDENTIFICATION_{sea\ ice})^c$
5: $WATER \leftarrow (PIXEL_{sea\ ice})^c$
6: $DISTRIBUTION \leftarrow$ floe size distribution
7: **return** $FLOE, BRASH, SLUSH, WATER, DISTRIBUTION$
Output: Classified layers and floe size distribution

It should be noted for the ice shape enhancement that some background (water) pixels may be turned into ice pixels by the operation of morphological cleaning. Those turned pixels should also be taken into account when determining the layers of slush and water, and when calculating the ice concentration. Therefore, the result of the ice pixel extraction should be updated after the step of ice shape enhancement by the union of previous ice pixel extraction result (by the k-means clustering method) with the individual ice piece identification result (by the ice shape enhancement algorithm).

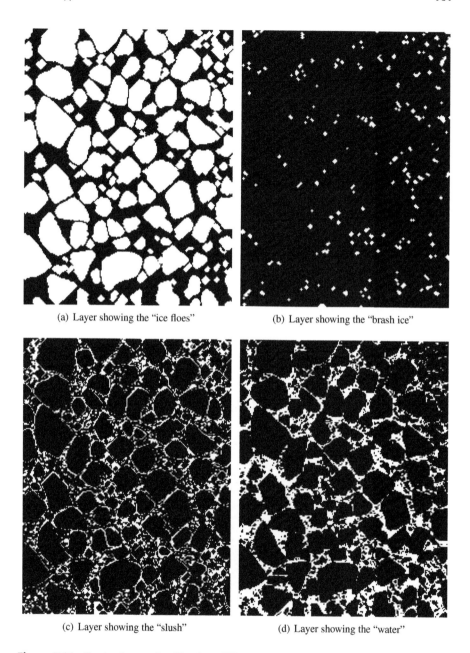

(a) Layer showing the "ice floes"

(b) Layer showing the "brash ice"

(c) Layer showing the "slush"

(d) Layer showing the "water"

Figure 7.14 Sea ice image classification of Figure 7.10(a) (Source: Figures from Q. Zhang and R. Skjetne, "Image Processing for Identification of Sea-Ice Floes and the Floe Size Distributions," *IEEE Transactions on Geoscience and Remote Sensing*, 53(5):2913-2924, 2015).

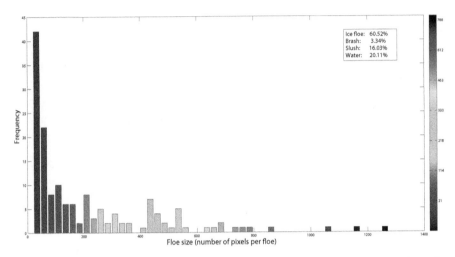

Figure 7.15 Floe size distribution histogram of Figure 7.14(a) (Source: Figure from Q. Zhang and R. Skjetne, "Image Processing for Identification of Sea-Ice Floes and the Floe Size Distributions," *IEEE Transactions on Geoscience and Remote Sensing*, 53(5):2913-2924, 2015).

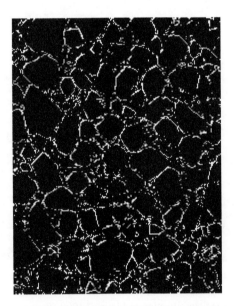

Figure 7.16 Residue ice (edge pixels) (Source: Figure from Q. Zhang and R. Skjetne, "Image Processing for Identification of Sea-Ice Floes and the Floe Size Distributions," *IEEE Transactions on Geoscience and Remote Sensing*, 53(5):2913-2924, 2015).

7.3 CASE STUDIES AND DISCUSSION

7.3.1 DISTORTED OVERALL SEA ICE IMAGE PROCESSING

Sea ice images usually cover a large area, and the illumination of the images is often non-uniform. Besides this, distortions may also exist in the image data distorted because of the oblique angles or optical lenses of the cameras. For example, aerial sea ice images usually have perspective distortion when the aerial vehicle orbits the observation field. Such issues will affect the final ice floe identification and data extraction results. This case study shows how to apply the proposed algorithms to process sea ice images. The main example is the image shown in Figure 7.17, which is covering a large area with perspective distortion.

Figure 7.17 Sea ice image with perspective distortion (Source: Figure from Q. Zhang and R. Skjetne, "Image Processing for Identification of Sea-Ice Floes and the Floe Size Distributions," *IEEE Transactions on Geoscience and Remote Sensing*, 53(5):2913-2924, 2015).

7.3.1.1 Local processing

Some ice information can be lost when globally extracting "light ice" and "dark ice" from a sea ice image in which non-uniform illumination or shadow problems exist. Moreover, a sea ice image typically contains multiple ice floes that crowd together, as shown in Figure 7.17, where parts of the floe boundaries become weaker than others. As discussed in Chapter 6, the GVF capture range derived by a uniform parameter sometimes cannot represent an overall ice image and should be adjusted according to each sub-image. Therefore, processing the local sub-images of the sea ice image is recommended to obtain an accurate segmentation result (but at the expense of more processing time and possibly manual intervention).

The image is first divided into smaller regions such that each region can be analyzed individually. To avoid image border effects as discussed in Section 6.5.3, it is recommended that the neighboring subregions overlap sufficiently. The ice edge detection algorithm (Algorithm 3) is then performed on each subregion to obtain a subsegmentated image. After that, the overlapping parts are removed and the subsegmented images are merged by an image stitching method. This procedure, illustrated in Figure 7.18, results in an overall segmented image.

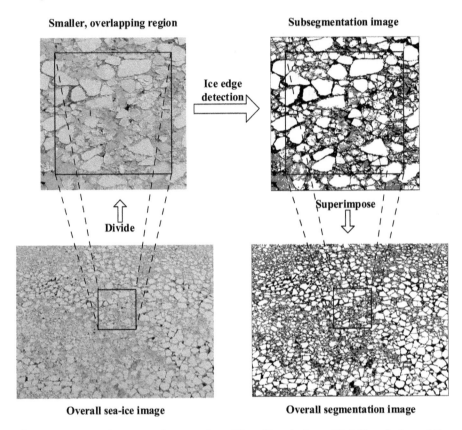

Figure 7.18 Local segmentation procedure. The white pixels are "light"ice pixels, and the gray pixels are "dark" ice pixels (Source: Figure from Q. Zhang and R. Skjetne, "Image Processing for Identification of Sea-Ice Floes and the Floe Size Distributions," *IEEE Transactions on Geoscience and Remote Sensing*, 53(5):2913-2924, 2015).

7.3.1.2   Geometric calibration

When perspective distortion exists in the image data, the final identification result, as illustrated in Figure 7.19, is not adequate for the calculation of size distribution statistics and other data. The ice floes in the far range of the image will seem smaller than those in the near range. This distortion will therefore induce errors in further analyses.

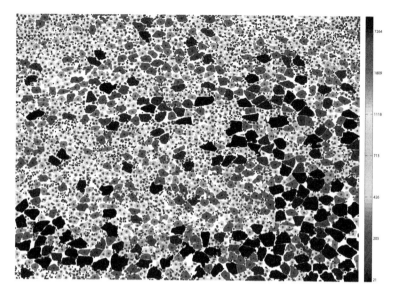

Figure 7.19 Ice floe and brash size distribution before orthorectification (Source: Figure from Q. Zhang and R. Skjetne, "Image Processing for Identification of Sea-Ice Floes and the Floe Size Distributions," *IEEE Transactions on Geoscience and Remote Sensing*, 53(5):2913-2924, 2015).

According to the method in Appendix A.1.1, the image can be orthorectified an-alytically when the values of the shooting angle and the field of view (FOV) of the camera are known[1]. This requires a sensor to measure the camera's shooting angle, such as an inertial measurement unit (IMU). For the particular image in Figure 7.17, the actual parameters of the camera were not measured. Hence, we have estimated the shooting angle to be approximately 20° and the FOV to be 46° (using the statis-tical similarities between the size distributions of near and far range of the image). Using this, we perform the orthorectification of the overall segmentation image.

To reduce the visual distortion caused by the fractional zoom calculation, the im-age will be enlarged, and the total number of pixels will increase after the orthorec-tification. The points between the pixels in the orthorectification coordinates that are mapped from the image coordinates must be interpolated. Each pixel holds quan-tized values that represent the color or gray scale of the image at a particular point. The technique of image interpolation, as introduced in Section 2.8, thus plays an important role in filling the values in those interpolated pixels by using known data to estimate values at unknown points. The nearest neighbor interpolation method is recommended for the geometric calibration. This is because each pixel is labeled

---

[1]Note that the rectification algorithm given by Appendix A.1.1 only considers the influence of pitch to give an easy example. In actual conditions, all influences from three degrees of freedom: pitch, roll, and yaw, must be considered.

with a distinct value indicating the catalogue ("light ice", "dark ice", or water) it belongs to after segmentation. If using other interpolation methods, new values, other than those three values for distinguishing "light ice", "dark ice", and water pixels, might be generated for the interpolated pixels.

For non-ridged and non-shielded sea ice images, when to execute the geometric calibration is important. On the one hand, the geometric calibration of the original sea ice image before running the ice identification algorithms may lead to a number of problems. First, the calibrated image, however, may be blurred because the values of the interpolated pixels are not the real values captured from the objects. As the number of interpolated pixels increase, the objects in the calibrated image become more blurry. The ice floe boundaries may then become weaker or even be lost, and the floe boundaries will become more difficult to be detected. If the geometric calibration is performed before the ice floe identification, the proposed algorithm may fail to detect the ice floes in the far range of the image because of their blurred boundaries. Second, due to the expansion of image size, more iterations are required for the calculation of the GVF force field to identify floe boundaries effectively, as discussed in Section 6.5.2, which is more time consuming. On the other hand, if the geometric calibration is carried out after the ice identification algorithms (sea ice shape enhancement Algorithm 4 or ice types classification Algorithm 5), some small ice floes located at the far end of the image would be mistaken for brash ice. Therefore, the geometric calibration should be performed on the segmented sea ice image (after Algorithm 3) before sea ice shape enhancement (Algorithm 4).

7.3.1.3 Result

We enhance the shapes of all the ice pieces (Algorithm 4) after orthorectification, and finally we obtain the ice floe and brash ice size distribution, as shown in Figure 7.20. Brash ice is dark blue, smaller floes are light blue, and larger floes are red. Brash positions are not shown, while the floe positions are denoted using a black dot.

A total of 2511 ice floes and 2624 brash ice pieces are identified from Figure 7.17. The coverage percentages are 65.98% ice floe, 5.03% brash ice, 17.52% slush, and 11.47% water. Instead of actual size of ice floe and brash (since we do not have the height above the sea-level for the camera for this example), the ice floe (brash) size is calculated by the number of pixels in the identified floe (brash). The relative ice floe distribution histogram is derived and shown in Figure 7.21, and the overall algorithm of the case study is concluded in Algorithm 6.

7.3.2 A PRELIMINARY SENSITIVITY STUDY

The number of iterations for the snake's evolution and for the GVF field are the most critical parameters that need to be tuned in this proposed sea ice image processing method. As of now, there is no explicit performance criterion for when a reasonable segmentation is achieved, other than the operators own judgment, and it is therefore very difficult to optimize the values of the parameters. Hence, a preliminary sensitivity study focusing on the snake iterations and the GVF field iterations is given here

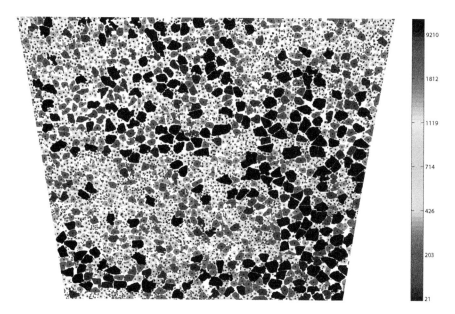

Figure 7.20 Ice floe and brash size distribution after orthorectification (Source: Figure from Q. Zhang and R. Skjetne, "Image Processing for Identification of Sea-Ice Floes and the Floe Size Distributions," *IEEE Transactions on Geoscience and Remote Sensing*, 53(5):2913-2924, 2015).

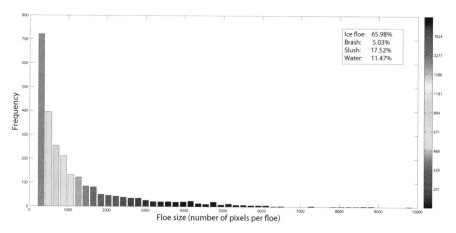

Figure 7.21 Ice floe size distribution histogram of Figure 7.20, excluding brash pieces (Source: Figure from Q. Zhang and R. Skjetne, "Image Processing for Identification of Sea-Ice Floes and the Floe Size Distributions," *IEEE Transactions on Geoscience and Remote Sensing*, 53(5):2913-2924, 2015).

Algorithm 6 Overall sea ice floe and brash identification algorithm

---

Input: Sea ice image
Start algorithm:
  1: $SUB \leftarrow$ sub-images divided from the input image
  2: **for** each sub-image $sub \in SUB$ **do**
  3:     $seg \leftarrow$ ice edge detection (Algorithm 3) on sub
  4: **end for**
  5: $SEG \leftarrow$ overall segmentation image with all seg stitched.
  6: $SEG \leftarrow$ geometric calibrated SEG
  7: $ID \leftarrow$ sea ice shape enhancement (Algorithm 4) on SEG
  8: $FLOE \leftarrow$ labeled regions in ID with large sizes.
  9: $BRASH \leftarrow$ labeled regions in ID with small sizes.
 10: $DISTRIBUTION \leftarrow$ floe size distribution
 11: **return** $FLOE, BRASH, DISTRIBUTION$
Output: Segmented layers and Floe size distribution

---

for further research on the automatic optimization of these two parameters.

This sensitivity study is carried on a 394×1038 sea ice image containing a mass of ice floes and brash ice, as seen in Figure 7.22, where the initial contours on the "light ice" are shown in Figure 7.23. The sensitivities of these two parameters are analyzed separately. In each analysis, one of the parameters is set to a fixed value while the other is varied from small to large, and the sea ice image processing method is performed several times. Then the resulting number of identified ice floes and brash ice pieces depending on the varied parameter are plotted and analyzed.

Figure 7.22   A marginal ice zone image used for the sensitivity study.

7.3.2.1   Snake's evolution iterations

As discussed in Section 6.5.1, an upper limit on the number of iterations can be set to stop the snake's evolution. In this sensitivity study, we vary the upper limit on snake iteration number from 1 to 122, while the GVF field is calculated by a fixed number

Figure 7.23   Initial contours on the "light ice" of Figure 7.22.

of 500 iterations (which is sufficiently large in this case). The relationship between the upper limit on snake iterations and the number of detected ice floes/brash ice pieces are shown in Figure 7.24.

When the upper limit on snake iterations is small, both over- and under-segmentations exist in the identification result. For example, if the snakes evolve only 1 iteration, which is far from enough for the snakes to find the floe/brash ice boundaries (Algorithm 3), the ice image is seriously over-segmented by the initial contours, as seen in Figure 7.25. However, most floes are still connected as a big "floe", containing several small segmented regions within. By applying the sea ice shape enhancement algorithm (Algorithm 4), these small regions inside the big "floe" are removed so that the floes become under-segmented, as shown in Figure 7.26(a).

By increasing the upper limit on snake iterations, the under-segmentation grad-ually decreases. This is because with more snake iterations, more floe/brash ice boundaries can be determined by the snakes (Algorithm 3), and more connected floes can thereby be separated such that these will not be merged by the sea ice shape en-hancement algorithm (Algorithm 4). However, if the upper limit is not large enough for the snakes to reach the floe boundaries, the over-segmentation will become dom-inant, as seen in Figure 7.26(b), where the result of 10 iterations for the snake is shown. This explains why the curves shown in Figure 7.24 increase progressively in the beginning.

As the upper limit increases, more and more snakes will reach the floe/brash ice boundaries, so that the over-segmentation and the number of detected ice floes and brash ice pieces decrease again. After the upper limit reaches a certain value where all the snakes are able to converge to their minima, which ideally are the desired floe/brash ice boundaries, the algorithm converges to a steady number of detected floes and brash pieces as seen in Figure 7.26(c) and in Figure 7.24.

7.3.2.2   GVF field iterations

In the sensitivity study on GVF field iterations, the upper limit on the snake's evo-lution is set to 100 iterations (which is large enough for the snakes to converge as

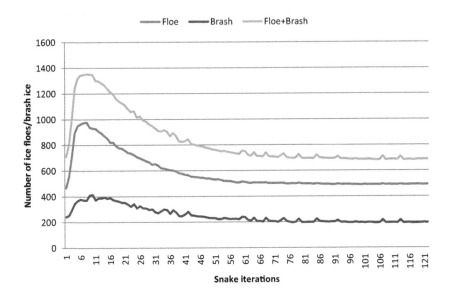

Figure 7.24 The number of detected ice floes/brash ice pieces based on the limit of the snake iterations.

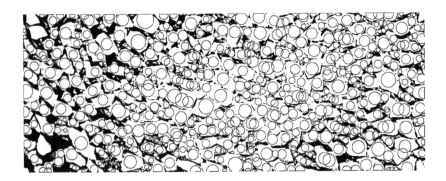

Figure 7.25 Segmentation result of Figure 7.22 when the snake evolves only 1 iteration.

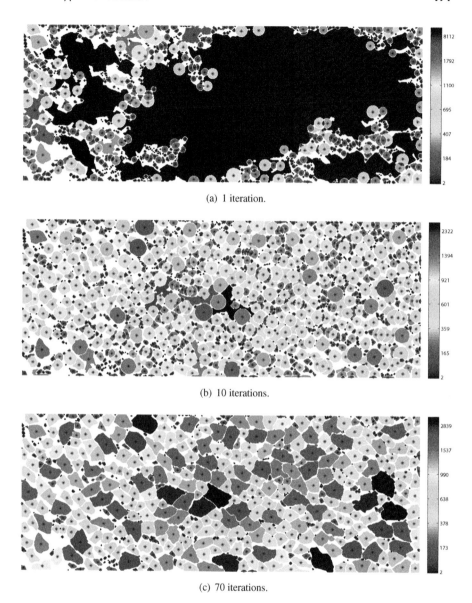

(a) 1 iteration.

(b) 10 iterations.

(c) 70 iterations.

Figure 7.26   Sea ice floe and brash ice identification results of Figure 7.22 based on different upper limits on snake iterations.

indicated by Figure 7.24). The iteration number for deriving the GVF field is then varied from 1 to 1500 at intervals of 20. The GVF field-dependent ice floes and brash ice curves are shown in Figure 7.27.

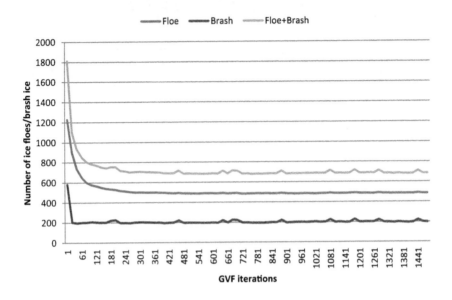

Figure 7.27 The GVF iteration-dependent ice floes/brash ice amount curves.

When the number of GVF field iterations is too small, the capture range of the external force fields generated by floe boundaries is limited such that their external forces can hardly act on the snakes. It is then difficult to attract the snakes to the floe boundaries. The external forces generated by the noise, on the other hand, are usually strong when the GVF capture range is low. These can easily guide the snakes to converge to the noise rather than the desired boundaries, leading to over-segmentation. Figure 7.28(a) shows this identification result when the GVF field is derived by only 1 iteration. It should be noted that the snakes now have enough iteration steps (100 steps) to converge to the minima. Most of these minima, however, are not the desired floe boundaries, but instead noise. Hence, most over-segmented ice pieces by the snakes (Algorithm 3) will in this case not be merged by the shape enhancement algorithm (Algorithm 4).

The more iterations used to calculate the GVF force fields, the larger the capture range becomes, and thus the more the snakes are able to converge to the correct floe boundaries. Furthermore, the external force vectors of the noise will eventually be overwhelmed by the vectors of the floe boundaries as the GVF field iterations increase. Therefore, the over-segmentation reduces gradually as shown in Figure 7.28(b).

The variation of the floe/brash ice sizes within the marginal ice zone image in

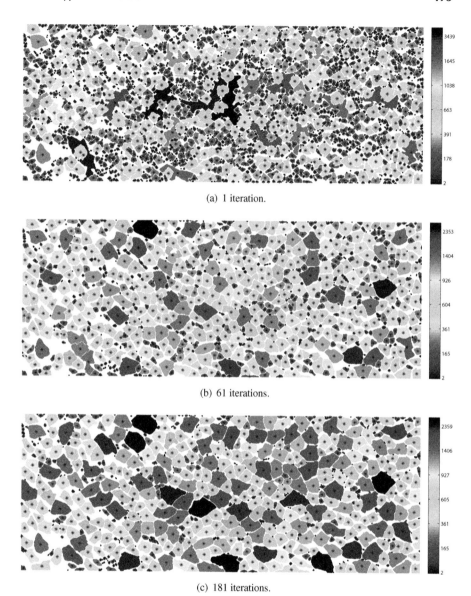

(a) 1 iteration.

(b) 61 iterations.

(c) 181 iterations.

Figure 7.28 Sea ice floe and brash ice identification results of Figure 7.22 based on different GVF field iterations.

Figure 7.22 is small. This means that the GVF fields for all the floe/brash pieces by the same number of iterations have similar extents of their capture ranges (relative to the floe sizes). So the snakes are able to find all floe/brash ice boundaries when the capture range is sufficiently large, as seen in Figure 7.28(c). This explains why the quantities of identified ice floes and brash ice tend to a steady value as the number of GVF field iterations become larger.

The boundaries of the brash ice are usually weaker than those of nearby ice floes. The identification of brash ice boundaries is therefore more sensitive to the parametric variations. The two parameter curves for the brash ice, shown in Figures 7.24 and 7.27, oscillate a little bit and are less smooth than those for the ice floes.

By performing such a sensitivity analysis, a pair of appropriate parameter values (i.e., the number of iterations for the snake's evolution and the iteration number for deriving the GVF field) can be estimated for a given sea ice image. Since the GVF iteration number should be set proportional to the image (object) size [61], the determined parameters may also be applied to other sea ice images of similar (object) scales.

8 Sea Ice Image Processing Applications

Several image processing techniques are introduced in the previous chapters, for sea ice observations. Some of them have been applied in Arctic offshore engineering for calculating the ice concentration and identifying individual ice floes from non-ridged sea ice images. The obtained results can be used for relating the ice field characteristics to the ice engineering problems, such as:

1. To quantify the efficiency of ice management for Arctic offshore drilling operations and automatically detect hazardous conditions, for example, by identifying large floes that escape the icebreakers operating upstream of the stationary drilling vessel. The size and shape of those floes, as identified by the image processing system, can be compared with the maximal allowed values, and a warning signal can be sent to the risk management system. Eventually, a decision to disconnect the floater might be taken, based on the identified operational ice conditions.

2. The managed ice concentration and ice floe sizes are essential parameters in the empirical formulas that estimate the ice loads on stationary Arctic marine structures [76, 116]. One of the largest concerns in ice management modeling is to accurately predicting not just the mean floe size resulting from an ice management system, but also the floe size distribution [15].

3. Individual ice floes identified by the image processing system can be used to initialize high-fidelity numerical models, such as those in Daley et al. [38], Vachon et al. [162], Sayed et al. [132], Sayed et al. [131], Gürtner et al. [51], Metrikin et al. [107], and Lubbad et al [98]. Individual snapshots of identified ice floes can, for instance, be used to validate the numerical models at various moments in time by matching the simulated ice fields with the actual ones.

4. The ice floe size and shape distribution, calculated from an identified ice field, can be used in synthetic ice field generators. These generators draw polygons from the distribution and use packing algorithms to place the polygons on a 2-dimensional plane. Such synthetic ice fields may be used to study various packing configurations with the same ice concentrations and floe size distributions as well as the variability of the resulting ice loads on a marine structure.

5. The identification of the ice field may provide an early warning of an ice compaction event, which can be dangerous if the ice-structure interaction mode changes from a "slurry flow"-type to a "pressured ice"-type, as defined by Wright [171] and discussed in Palmer and Croasdale [116].

6. Finally, the ice-drift speed and direction (velocity) can be estimated by applying an image analysis to sequential frames. The ice-drift velocity is an important parameter for ice management, because it poses requirements on the speed and number of icebreaking vessels and may indicate an approaching ice drift reversal

scenario (which usually happens when the ice drift tends to zero velocity).

In this chapter, the applications of the image processing methods for identifying sea ice parameters are presented and discussed.

8.1 A SHIPBORNE CAMERA SYSTEM TO ACQUIRE SEA ICE CONCENTRATION AT ENGINEERING SCALE

The expedition Oden Arctic Technology Research Cruise 2015 (OATRC'15) was carried out by the Norwegian University of Science and Technology (NTNU) in collaboration with the Swedish Polar Research Secretariat (SPRS) [97]. Two icebreakers, Oden and Frej, were employed during this research cruise into the Arctic Ocean. Among many research activities, the extraction of sea ice information is of great importance. Various technologies and equipment suited to different needs/scales were available during the cruise to probe and document ice conditions.

Sea ice concentration is an important parameter for the calculations of ice actions and its effect on Arctic marine structures and for the evaluation of icebreaker performance. Various methods exist nowadays to monitor these parameters, ranging from geophysical scale to local scale. During the OATRC'15, an Ice Concentration camera was installed and corresponding algorithms were developed to achieve real-time quantification of ice concentration [94].

8.1.1 INSTALLATION OF CAMERAS

The installation location of the Ice Concentration camera, and its viewing directions, are illustrated in Figure 8.1. The Ice Concentration camera was composed of four lenses, and four images were correspondingly produced for each shot. These cover a wide horizontal view angle of about 180°. Out of the four lenses, we managed to calibrate the middle two, and the ice concentration analysis is thus based on these two calibrated lenses. The images are captured at 1 frame per 10 seconds.

8.1.2 METHODS

The Ice Concentration camera is looking obliquely toward the ship transiting direction. The obtained image is therefore distorted with more/less pixels in the near/far field. A perspective rectification process is needed to project the originally captured image onto the sea plane. Figure 8.2 schematically illustrates how the original image from the starboard lens is projected back into the real ice field via perspective orthorectification. Appendix A.1.1 gives an analytical method for the perspective orthorectification. However, for obtaining ice concentration in real time, a fast perspective orthorectification method is necessary. Thus, a linear rectification [3, 57], as illustrated in Appendix A.1.2, is adopted in this analysis.

To conduct the linear rectification, only the relative physical coordinates (denoted as $x'y'$-coordinate system) of four corners of a rectangular and their corresponding coordinates in the image (denoted as xy-coordinate system) are needed to establish a transformation matrix. The relative physical coordinates are given by the size of

Figure 8.1 Locations of the installed cameras on IB Frej (Source: Figure from W. Lu, Q. Zhang, R. Lubbad, S. Løset and R. Skjetne, "A Shipborne Measurement System to Acquire Sea Ice Thickness and Concentration at Engineering Scale," In *Proceedings of OTC Arctic Technology Conference*, St. John's Newfoundland and Labrador, Canada, 2016).

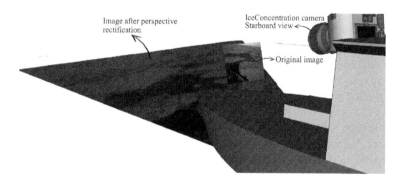

Figure 8.2 Perspective rectification process and physical size of the analyzed region (starboard image as an example) (Source: Figure from W. Lu, Q. Zhang, R. Lubbad, S. Løset and R. Skjetne, "A Shipborne Measurement System to Acquire Sea Ice Thickness and Concentration at Engineering Scale," In *Proceedings of OTC Arctic Technology Conference*, St. John's Newfoundland and Labrador, Canada, 2016).

a checkerboard for calibration (see Figure 8.3). The camera is fixed at a height of 25.50 m over sea plane, and the checkerboard is positioned at a height of 5.90 m over sea plane on the main deck. Therefore, the obtained image in the $x'y'$-plane needs to be further projected onto the sea plane. Figure 8.3 illustrates the images we used to calibrate the lens on the port and starboard side. The obtained transformation matrix is then used for all the obtained images, where we note that the influence of the ship motion on the transformation has been neglected.

(a) Lens towards the port side.   (b) Lens towards the starboard side.

Figure 8.3 Calibration of the Ice Concentration camera lens (note that the calibration pictures are taken at different time) (Source: Figure from W. Lu, Q. Zhang, R. Lubbad, S. Løset and R. Skjetne, "A Shipborne Measurement System to Acquire Sea Ice Thickness and Concentration at Engineering Scale," In *Proceedings of OTC Arctic Technology Conference*, St. John's Newfoundland and Labrador, Canada, 2016).

Different factors, such as installation height, focal length, and shooting angle of the camera, could all contribute to the perspective rectification errors [92]. The shooting angle of the camera and the pixel's y location in the xy-plane significantly influence the perspective rectification error. Intuitively, this reflects the fact that the objects that are visually closer to the camera in the image can be more accurately quantified. Lu and Li [92] set up a limitation to the camera's shooting angle such that no pixels above horizon should be included; they also discussed the shielding effects that occur when the objects are too far away. Considering all these potential error sources, we restrict our analysis only on the sea/ice features that are in the vicinity of the ship. Figure 8.4(a) gives an example of the original images captured by the port and starboard side lenses. In these images, two areas within the white triangles are cropped away, and only the remaining pixels are assumed to be close enough such that no major error could be induced after the perspective rectification.

Figure 8.4(b) illustrates the images after perspective rectification (via bilinear interpolation) and its physical dimensions in the sea plane. We note that the port and starboard side camera views are not symmetric due to challenging precision control in the field installation. Nevertheless, we are analyzing a region that is approximately 400 m by 100 m for extraction of ice concentration information.

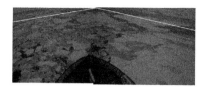

(a) Original images.

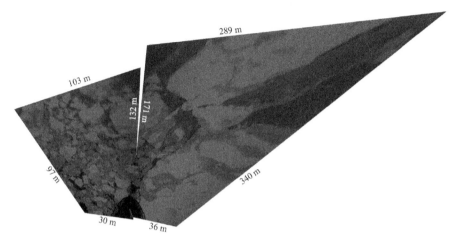

(b) Rectified images.

Figure 8.4 An example showing the size of the analyzed region (engineering scale) (Source: Figure from W. Lu, Q. Zhang, R. Lubbad, S. Løset and R. Skjetne, "A Shipborne Measurement System to Acquire Sea Ice Thickness and Concentration at Engineering Scale," In *Proceedings of OTC Arctic Technology Conference*, St. John's Newfoundland and Labrador, Canada, 2016).

In order to extract the pixels that represent ice and those that represent open water, both the global Otsu thresholding method and the k-means clustering method are utilized to separate different regions based on the grayscale of the pixels. In the Otsu thresholding method, the identified threshold values enable us to turn a grayscale ice image into a binary image with black water and white ice. In the k-means clustering method, the ice image is categorized into more regions. Particularly, we identify three clusters ($k = 3$) for the k-means clustering method by assuming the original image being composed of water (black region), wet ice (gray region, generally composed of ice rubbles, young ice, and melt ponds), and dry ice (white region). It is worth mentioning that if we choose two clusters for the k-means method ($k = 2$), then this method is approximately reduced to the Otsu thresholding method. After the separation of different regions, the calculation of ice concentration is reduced to calculate the pixel number of the ice regions over the total pixel number. The only nontrivial task here is to ascertain which regions can be categorized as ice. For the global Otsu method, it is straightforward to define white region as ice. However, this is overly simplified in order to cope with the complicated ice environment. As shown in Figure 8.4, some gray areas are actually melt ponds or water on ice and should be included as ice. Therefore, in the forthcoming calculation by the k-means ($k = 3$) method, we include both the white and gray region as ice.

8.1.3   RESULTS AND DISCUSSIONS

Based on the methods introduced above, we selected a time window from Sep-26-2015 08:00 to 14:00. The ice concentration history by both the k-means ($k = 3$) method and the global Otsu method are shown in Figure 8.5.

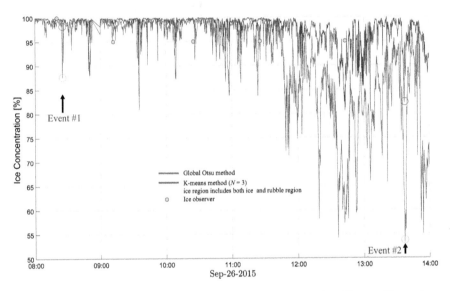

Figure 8.5   Ice concentration variations in the analyzed time window (Source: Figure from W. Lu, Q. Zhang, R. Lubbad, S. Løset and R. Skjetne, "A Shipborne Measurement System to Acquire Sea Ice Thickness and Concentration at Engineering Scale," In *Proceedings of OTC Arctic Technology Conference*, St. John's Newfoundland and Labrador, Canada, 2016).

Figure 8.5 shows that the k-means method generally yields a slightly higher concentration in comparison to the global Otsu method. This is understandable because some ice rubble/melt ponds area (in gray) is also included as ice. In general, these two methods predict similar trend on ice concentrations variation in comparison to onboard ice observers inputs. However, since the ice surface is scattered with melt ponds, the global Otsu thresholding method typically mistakes these melt ponds as open water and thus under-estimates the ice concentration. In addition to the general trends, we consider two special events (Event #1 and #2 in Figure 8.5), where large discrepancies between these two methods are identified. These two extreme events are depicted by the original pictures and the converted binary images as shown in

Figure 8.6. In Event #1, a large part of young ice of darker color was categorized as ice region by the k-means clustering method, while it was categorized as open water by the global Otsu thresholding method. In Event #2, due to the adverse illumination condition, a large less light intensive area on the ice surface was mistakenly assigned as open water by the global Otsu thresholding method. On the contrary, the k-means clustering method suffers less from this adverse illumination, and less shadow was categorized as open water. Comparing these two methods, the k-means clustering method demonstrates a better flexibility in categorizing a grayscale image into different regions, and one can have a better control on what to include in the ice region for the calculation.

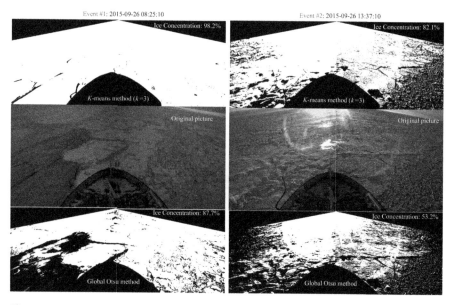

Figure 8.6 Comparison of ice concentration calculation methods in the two selected events (Source: Figure from W. Lu, Q. Zhang, R. Lubbad, S. Løset and R. Skjetne, "A Shipborne Measurement System to Acquire Sea Ice Thickness and Concentration at Engineering Scale," In *Proceedings of OTC Arctic Technology Conference*, St. John's Newfoundland and Labrador, Canada, 2016).

8.2 NUMERICAL CHARACTERIZATION OF A REAL ICE FIELD FOR PARAMETRIZATION OF AN ICE SIMULATOR

A remote sensing mission to determine ice conditions was performed by the Northern Research Institute (NORUT) at $78°55'N$ $11°56'E$, Hamnerabben, Ny-Ålesund from May 6th to 8th, 2011. The objectives of this mission were to observe and learn from UAV (unmanned aerial vehicle) operation in the Arctic and obtain remotely sensed data of sea ice features from a mobile sensor platform.

A CryoWing UAV, as shown in Figure 8.7, was used as a mobile sensor platform for the mission. This UAV was designed for cryospheric measurements and environmental monitoring. It has flexibility in coverage and in spatial and temporal resolution, which are three important sensor-platform attributes. The technical specification of the CryoWing is found in Table 8.1(a). The basic instrumentation of the CryoWing is an onboard computer that controls the different payload instruments, stores data to a solid-state disk, and relays data to the ground. The onboard payload system has a GPS receiver and a 3-axis orientation sensor independent of the avionics system. The sensor device used in this analysis is a digital visible camera with specifications found in Table 8.1(b).

Figure 8.7 The CryoWing UAV operation at Ny-Ålesund (Source: Figure from Q. Zhang and R. Skjetne, "Image Processing for Identification of Sea-Ice Floes and the Floe Size Distributions," *IEEE Transactions on Geoscience and Remote Sensing*, 53(5):2913-2924, 2015).

The UAV flew in the inner part of Kongsfjorden close to a buoy that had been deployed on the ice cover to collect high-resolution images of sea ice, and several image processing algorithms have then been applied to these images to extract useful information of the sea ice, such as ice concentration, ice floe boundaries, and ice types.

Figure 8.8 shows an aerial image of the MIZ obtained in the UAV mission. Since the image is undistorted, Algorithms 3, 4, and 5 are carried out directly to this image. The resulting four layers of a sea ice image and their coverage percentages are 76.73% ice floe shown in Figure 8.9(a), 0.46% brash ice shown in Figure 8.9(b), 9.05% slush shown in Figure 8.9(c), and 13.76% water shown in Figure 8.9(d). Based on the four layers, a total of 498 ice floes and 201 brash ice pieces are identified from Figure 8.8.

The ice floe/brash ice piece size are determined by the number of pixels in the identified floe/brash ice piece. The ice floe and brash ice are labeled in different colors based on their sizes with '*' and '·' at their centers, as shown in Figure 8.10. The resulting ice floe size distribution histogram is then shown in Figure 8.11.

Table 8.1
Specifications of UAV and camera.

(a) CryoWing technical specifications

Weight	30 kg max take off weight
Wingspan	3.8 m
Cruise speed	100 – 120 km/h
Range/endurance	500 km / 5 h
Max altitude	2500 m dynamic range, 5000 m absolute
Payload capacity	Max 15 kg including fuel load
Engine	Two stroke gasoline
Navigation	GPS
Ground equipment	PC with modem, RC control
Flight	Autonomous, but under ground control
Communication	GSM or Iridium satellite modem

(b) Visible spectrum camera specifications

Camera model	Canon EOS 450D
Lens type	Canon EF 28 mm f/2.8
Aperture value	11.00
Sensor	22.2×14.8 mm CMOS
ISO	200
Dimensions	4290×2856
Resolution	960 dpi
Exposure time	$1/250$ sec
Sampling frequency	0.66 Hz

Source: Tables from Q. Zhang and R. Skjetne, "Image Processing for Identification of Sea-Ice Floes and the Floe Size Distributions," *IEEE Transactions on Geoscience and Remote Sensing*, 53(5):2913-2924, 2015.

Figure 8.8 An aerial image of the marginal ice zone obtained in the UAV mission (Source: Figure from Q. Zhang and R. Skjetne, "Image Techniques for Identifying Sea-Ice Parameters," *Modeling, Identification and Control*, 35(4):293-301, 2014).

8.2.1 SEA ICE NUMERICAL MODELING

Ice floes segmented in Figure 8.9(a) are not necessarily convex. In the MIZ, however, the ice floes generally exhibit a rounded shape [129]. To better approximate the geometry of the ice floes, and also for numerical representation, the ice floe geometries are further simplified by bounding minimum-area-polygons, while the brash ice pieces are reshaped by disks of equivalent areas.

Figure 8.12 shows an example of the sea ice modeling for Figure 8.10. A close-up view of a few ice floes and brash ice pieces in the middle of Figure 8.10 is given in Figure 8.13. The boundaries of the fitted polygons (circles) are superimposed into Figure 8.12. The centers of the identified ice floes (brash ice pieces) are marked with '∗ (·)', while the centers of the polygonized floes (circularized brash ice) are marked with red '+ (·)'.

Figure 8.13 shows how the polygonization (circularization) modifies the floes (brash ice). We note that the polygonized floes will not be smaller than the actual identified floes, and they may overlap other floes and brash ice pieces. Identifying the overlaps between the ice floes or brash ice pieces may be important when using the resulting ice field characterization as an initial condition of a numerical simulation with a 3-dimensional capacity. To accommodate such a need, an "overlap flag" of each floe-floe and floe-brash is added to each polygonized floe to indicate the serial number of which floes and brash ice pieces the current floe overlaps with.

Figure 8.14 presents the colorized histogram of the updated ice field and the polygonized floe size distribution of Figure 8.12(a). By subtracting the histograms of Figure 8.12(a) from Figure 8.10, we get the floe size distribution error, as seen in the histogram in Figure 8.15, due to the shape simplification of the ice pieces. This gives us a quantification of the change of ice floes of various size.

The processed results are stored as a MATLAB® data structure, within which all necessary information for numerical modeling is available. Detailed information of the data structure is presented in Appendix B. The ice image data structure has been used to generate a sea ice field for numerical simulation of ice-structure interaction. Figure 8.16(a) shows the initial ice field condition when importing its numerical

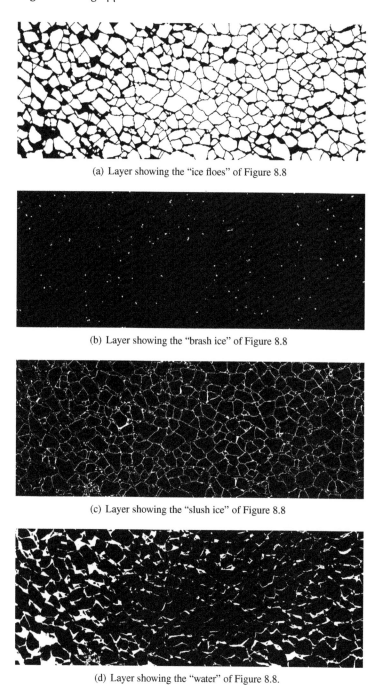

(a) Layer showing the "ice floes" of Figure 8.8

(b) Layer showing the "brash ice" of Figure 8.8

(c) Layer showing the "slush ice" of Figure 8.8

(d) Layer showing the "water" of Figure 8.8.

Figure 8.9 Identification result producing four layers for Figure 8.8 (Source: Figure from Q. Zhang and R. Skjetne, "Image Techniques for Identifying Sea-Ice Parameters," *Modeling, Identification and Control*, 35(4):293-301, 2014).

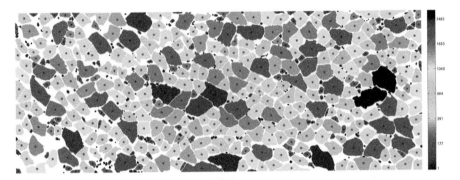

Figure 8.10   Floe and brash ice size distribution for Figure 8.8 (Source: Figure from Q. Zhang and R. Skjetne, "Image Techniques for Identifying Sea-Ice Parameters," *Modeling, Identification and Control*, 35(4):293-301, 2014).

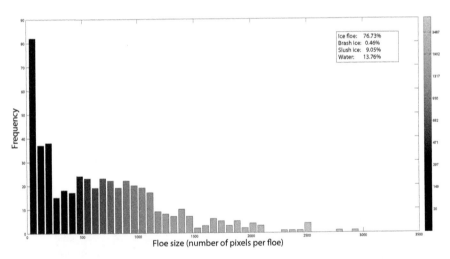

Figure 8.11   Floe size distribution histogram obtained from Figure 8.10 (Source: Figure from Q. Zhang and R. Skjetne, "Image Techniques for Identifying Sea-Ice Parameters," *Modeling, Identification and Control*, 35(4):293-301, 2014).

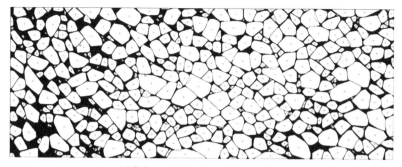

(a) Polygonized ice floes in Figure 8.10.

(b) Brash ice simplified by size equivalent disks in the domain of Figure 8.10.

Figure 8.12 Ice field characterization for Figure 8.10.

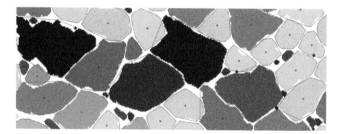

Figure 8.13 A close-up view of floe ice, brash ice, and their corresponding simplifications in the modeling. The closed curves are the boundaries of the modified floes (brash ice). '* (·)' are the centers of the identified ice floes (brash ice pieces), and '+ (·)' are the centers of the polygonized floes (simplified brash ice).

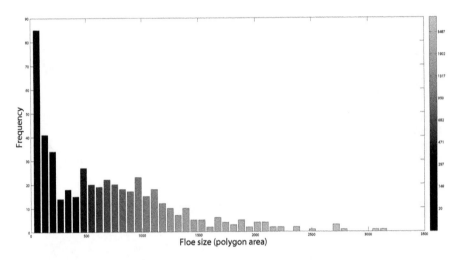

Figure 8.14   Polygonized ice floe size distribution histogram.

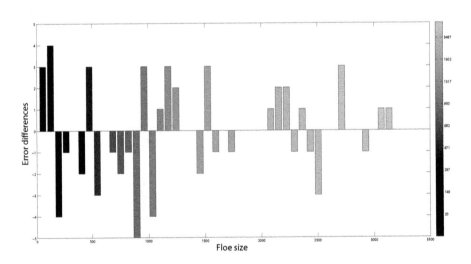

Figure 8.15   Error between the floe size distributions of Figure 8.11 and Figure 8.14.

representation into a non-smooth discrete element method (DEM) based simulator [38, 96, 106]. The ice floe/brash ice pairs with overlap are labeled with dark-gray color, while the others are labeled with light-gray color. Figure 8.16(b) shows the final generated ice field after identifying and resolving the overlaps by collision detection and collision response calculations [36].

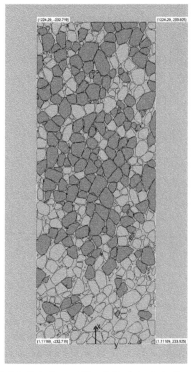 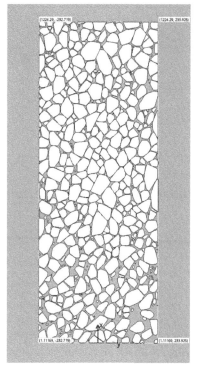

(a) Initial phase of ice field characterization with overlaps. light-gray: floes without overlaps; dark-gray: floes with overlaps.

(b) Final ice field characterization without overlaps.

Figure 8.16   Synthetic ice field generation according to Figure 8.12.

8.3   SEA ICE FLOE SIZE STATISTIC

During the field activities in the OATRC'15 expedition, an AS-335NP helicopter based on icebreaker (IB) Oden was employed to shoot images over the ice field with a camera system to document the ice management operations. The camera system consisted of a camera support with a 6-axis gyro stabilizer and a Red Dragon camera with Fujinon $25 - 300$ mm lens. The camera was installed on the support, as seen in Figure 8.17, and its specifications can be found in Table 8.2.

One of the helicopter flight missions was to capture ice conditions in the MIZ, and

—A 6-axis gyro stabilized
   camera support;
—Filming height:
   around 1314 m;
—Flying speed: 40 m/s;
—Filming rate: 25 fps;
—Total coverage:
   3 by 3 nm

Figure 8.17 Helicopter camera.

Table 8.2
Specifications of the visible spectrum camera mounted on the helicopter.

Lens type	Fujion 35mm.
Focal length	28.5 mm.
Dimensions	5568 × 3132.
Sampling frequency	0.1 Hz.

much valuable ice data of the MIZ is contained in the captured images. The objective of our algorithms presented in this book is to extract and quantify this information. An example of an aerial sea ice image over the MIZ can be seen in Figure 8.18. By applying the image processing algorithms to Figure 8.18, a total of 2888 ice floes and 3452 brash ice pieces are identified. The coverage percentages of the different sea ice types are: 58.00% ice floes, 4.85% brash ice, 21.21% slush, and the remaining 15.94% is water.

Taking IB Oden, whose physical size is known, as a reference, the physical sizes of the identified ice floes can be estimated by comparing the sizes of the floes and IB Oden in the images obtained from the heli-borne camera at equivalent heights. As mentioned in Section 1.2.3, ice floe sizes are practically presented by the "representative diameters". The sizes of the identified ice floes are represented based on their mean clipper diameter (MCD) sizes, defined by:

$$L_i = \sqrt{\frac{4A_i}{\pi}} \tag{8.1}$$

where A_i is the floe area. The ice floes identification result of Figure 8.18 is presented in Figure 8.19, where the identified ice floes, marked with white dots at centers, are labeled in different colors based on their MCD sizes for a better visualization. In addition, the ice floe size distribution histogram is shown in Figure 8.20.

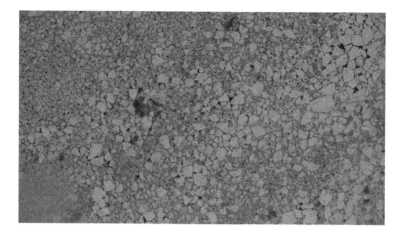

Figure 8.18   An aerial sea ice image captured during the OATRC'15 expedition.

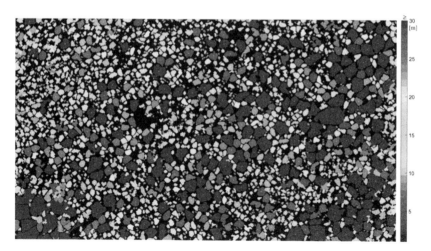

Figure 8.19   Ice floe identification result of Figure 8.18. The white dots are the centers of the identified ice floes, and the color bar shows the MCD of the ice floes.

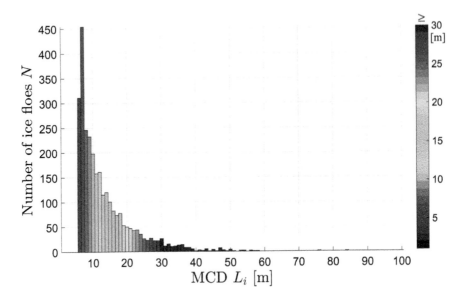

Figure 8.20   Ice floe size distribution histogram of Figure 8.19.

To characterize the sea ice floe sizes in the MIZ, the floe size distribution is typically expressed as the cumulative floe size distribution $N_c(L)$, defined as the number of ice floes per unit area with size no smaller than L:

$$N_c(L) = \frac{N(\geq L)}{N_{total}} \tag{8.2}$$

where $N(\geq L)$ is the number of floes with size larger than or equal to L, and N_{total} is the total number of floes in the sampling area. The obtained cumulative floe size distribution is reported to follow a power law [129, 157]:

$$N_c(L) \propto L^{-\alpha} \tag{8.3}$$

The exponent α in Equation 8.3 is the slope of the power law curve on log-log plot (which is a straight line), and it is estimated to be 1.3704 by fitting the power law distribution to the cumulative floe size distribution of Figure 8.19. The fitting result is shown in Figure 8.21, showing a good fit for the smaller floe sizes.

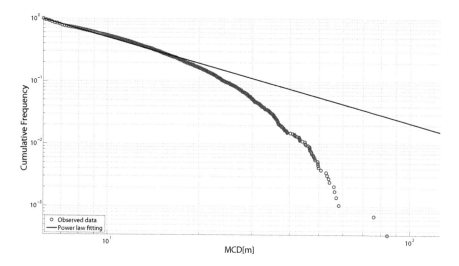

Figure 8.21 The cumulative floe size distribution of Figure 8.19 and its power law fitting.

9 Model Sea Ice Image Processing Applications

The ice concentration and floe size distribution are important ice parameters in ice-structure analyses. Before performing an analysis in full scale, model-scale experiments are typically conducted to verify the structural engineering design and to parameterize numerical models and equations. In ice engineering, this is done in an ice model basin, where sea ice dimensions and properties are scaled down using a sophisticated freezing process and chemicals to ensure as accurate representation as possible of the sea ice in its smaller model scale.

In this chapter, we refer to a series of model-scale experiments that were conducted for Dynamic Positioning (DP) of ship-shaped floaters in ice. These were carried out in the ice tank at the Hamburg Ship Model Basin (HSVA) in the summer of 2011 [68]. In these experiments, the behavior of two different vessels in a broken-ice field were studied, and it allowed for the testing of relevant image processing algorithms. In particular, image processing techniques are applied to derive the ice concentration and floe size distribution in the model basin. Several points in time are analyzed in order to describe the evolution of the ice field. The applied techniques include the methods for calculating the ice concentration and identifying individual ice floes in the vicinity of the model vessel.

9.1 EXPERIMENTAL SETUP AND MODEL SEA ICE IMAGE DATA

In the research project "Dynamic Positioning in Ice Covered Waters (DYPIC)" [52], two different model vessels have been tested at HSVA — an Arctic drillship and a polar research vessel. Each vessel was tested both in free running and oblique towing configurations. For image processing, the analysis is limited to the drillship in the oblique towing mode, based on the test campaign conducted in May 2011. Different heading and velocity profiles were tested. In the analyzed runs, the heading was constant at 180° and the velocity of the towing carriage with the model was increased halfway. By doing this, the full-scale ice-drift velocity of 0.25 knots was simulated in the first part of the test and 0.50 knots in the second part.

A managed ice condition was obtained by cutting the level ice layer into predefined ice floe shapes. Four different types of ice fields were tested, varying in ice concentration and ice floe size distribution, as shown in Table 9.1. The runs were sequentially executed, starting with run no. 5100. This initial ice field was prepared by cutting a 54-meter long ice sheet into pieces and distributing them over 64 meters of the tank length.

The cutting procedure was as follows. First, several strips of ice were cut in the longitudinal direction of the ice tank. One strip of 1.50 m width, four strips of 1.00

m width, and nine strips of 0.50 m width coincide with the percentages recorded in Table 9.1. Next, these strips were cut off such that the length was equal to the width of the strip, resulting in square ice floes. For run no. 5200, a number of floes were taken out of the basin in order to reduce the ice concentration. For run no. 5300, all present ice floes were cut into half diagonally. Finally, the removed ice floes were reinserted, but cut in half, in run no. 5400.

Table 9.1

Managed ice conditions in the test runs, target values (model-scale).

Run no.	Ice Concentration [%]	Floe size 1 (45%) [m]	Floe size 2 (40%) [m]	Floe size 3 (15%) [m]
5100	86	0.50	1.00	1.50
5200	70	0.50	1.00	1.50
5300	70	0.25	0.50	0.75
5400	86	0.25	0.50	0.75

Source: Table from Q. Zhang, S. van der Werff, I. Metrikin, S. Løset and R. Skjetne, " Image Processing for the Analysis of an Evolving Broken-Ice Field in Model Testing," In *ASME 31st International Conference on Ocean, Offshore and Arctic Engineering*, Rio de Janeiro, Brazil, 2012.

Ice conditions were captured by several means. First, a top-view camera was used before each test run to take 28 pictures over the total ice-covered basin. Stitching these photos together resulted in a complete overview of the ice floe distribution in the ice tank. Second, a top-view video camera moving along with the carriage and model was used to capture the local conditions around the model vessel during each run. Other video cameras were installed as well in order to investigate the behavior of the model vessel in the broken ice and its interaction with the ice. Two cameras were mounted on the main carriage or the service carriage to capture the vessel from the side and front. A fish-eye camera on the main towing carriage was able to film the vessel from above. Photographs were taken manually during the tests.

9.2   ICE CONCENTRATION

Ice concentration has been identified as one of the most influential parameters on the magnitude of experienced forces during model tests [164, 32]. Otsu thresholding and k-means clustering methods, as introduced in Chapter 3, are applied to the model sea ice image/video data to derive the ice concentration.

9.2.1   ICE CONCENTRATION DERIVING FROM OVERALL TANK IMAGE

After preparing the ice field and before the test run started, a top-view camera was positioned over the total ice-covered basin to produce an overall image of the complete ice field, for instance the total image in Figure 9.1 shows the distribution of ice floes over the tank length in run no. 5100, and Figure 9.2 shows its grayscale histogram.

Figure 9.1   Overall tank image for run no. 5100. Target ice concentration 86% (Source: Figure from Q. Zhang, S. van der Werff, I. Metrikin, S. Løset and R. Skjetne, " Image Processing for the Analysis of an Evolving Broken-Ice Field in Model Testing," In *ASME 31st International Conference on Ocean, Offshore and Arctic Engineering*, Rio de Janeiro, Brazil, 2012).

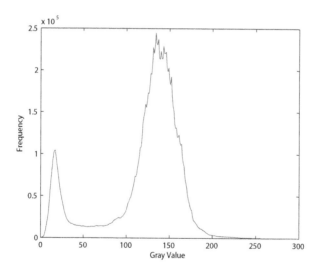

Figure 9.2   Histogram of the overall tank image for run no. 5100 (Source: Figure from Q. Zhang, S. van der Werff, I. Metrikin, S. Løset and R. Skjetne, " Image Processing for the Analysis of an Evolving Broken-Ice Field in Model Testing," In *ASME 31st International Conference on Ocean, Offshore and Arctic Engineering*, Rio de Janeiro, Brazil, 2012).

The ice floes were segmented from the water by applying the global Otsu, local Otsu, and *k*-means methods. The ice concentrations were calculated individually based on these three methods. The results can be found in Figures 9.3, 9.4, and 9.5.

Figure 9.3 Run no. 5100. Global Otsu method, $IC = 83.17\%$, threshold $= 84$ (Source: Figure from Q. Zhang, S. van der Werff, I. Metrikin, S. Løset and R. Skjetne, " Image Processing for the Analysis of an Evolving Broken-Ice Field in Model Testing," In *ASME 31st International Conference on Ocean, Offshore and Arctic Engineering*, Rio de Janeiro, Brazil, 2012).

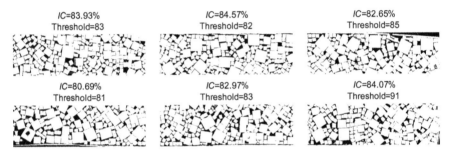

Figure 9.4 Run no. 5100. Local Otsu method, average $IC = 83.14\%$, average threshold $= 84$ (Source: Figure from Q. Zhang, S. van der Werff, I. Metrikin, S. Løset and R. Skjetne, " Image Processing for the Analysis of an Evolving Broken-Ice Field in Model Testing," In *ASME 31st International Conference on Ocean, Offshore and Arctic Engineering*, Rio de Janeiro, Brazil, 2012).

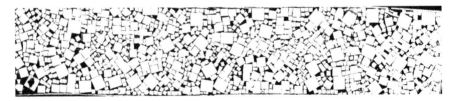

Figure 9.5 Run no. 5100. *K*-means method, 2 clusters, $IC = 82.86\%$ (Source: Figure from Q. Zhang, S. van der Werff, I. Metrikin, S. Løset and R. Skjetne, " Image Processing for the Analysis of an Evolving Broken-Ice Field in Model Testing," In *ASME 31st International Conference on Ocean, Offshore and Arctic Engineering*, Rio de Janeiro, Brazil, 2012).

The grayscale histograms of the overall tank images are clearly bimodal as seen in Figure 9.2. Moreover, the illumination of the overall tank image is almost uniform and only one type of ice existed in the tank. It means that both assumptions of the global Otsu thresholding method hold true. Hence, the differences in the calculated results between the global and local Otsu methods are small. Furthermore, the k-means method demonstrates results that are very close to the Otsu thresholding methods. Both of the methods are effective in this case.

The results of the ice concentration analysis were compared with the target ice concentration values. The results are presented in Table 9.2. The ice concentrations derived from the different methods are approximately $3 - 8\%$ smaller than the targeted value. A source of error is the upper right corner of the image that sits outside the tank. However, the main reason is believed to be imperfect ice sheet preparation, where a portion of the ice sheet was lost during the ice redistribution. This led to decreased ice concentration compared to the target values.

Table 9.2

Ice concentrations derived from different methods.

Methods	Target Value	Global Otsu	Local Otsu	K-means
Run no. 5100	86%	83.17%	83.14%	82.86%
Run no. 5200	70%	62.50%	62.51%	62.00%

Source: Table from Q. Zhang, S. van der Werff, I. Metrikin, S. Løset and R. Skjetne, " Image Processing for the Analysis of an Evolving Broken-Ice Field in Model Testing," In *ASME 31st International Conference on Ocean, Offshore and Arctic Engineering*, Rio de Janeiro, Brazil, 2012.

9.2.2 ICE CONCENTRATION DERIVING FROM MODEL SEA ICE VIDEO

A video is composed from a sequence of frames. The motions captured by the video are retrieved by analyzing a number of frames. The time variation of the ice concentration can be evaluated by plotting the individual frame analysis results over time.

The four videos supplied by HSVA are more than 24 minutes long with a frame rate of 25 fps. These videos are used to evaluate the near-vessel ice concentration over time. Before applying the algorithms to these videos, one frame per second is found sufficient, and each frame was fed to the program for further processing.

The distortion caused by the fisheye camera will make the scale in the middle of the images/videos larger than the circumambience (an analytical and an approximation method for calibrating fisheye distortion can be found in Appendix A.2). This phenomenon has an insignificant effect on the boundary detection, but may have some influence on the analyses of the ice concentration and ice floe sizes. In our cases, the fisheye distortion existing in the videos is insignificant, therefore can be negligible.

The light sources in the ice tank were reflected by the water and the ice. Due to the bright characteristics of the lights, they may be identified as ice floes by the algorithm, and consequently, the ice concentration may be estimated slightly too high.

The impediments around the tank are removed, and the vessel in the middle bottom of the tank is eliminated by a black rectangle (see Figure 9.6). The vessel box removed from the images may slightly affect the quality of the results for ice concentration.

(a) One frame in the original video.   (b) Domain image.

Figure 9.6   Original frame in the video and pre-processed frame. Run no. 5100.

The global Otsu and the k-means clustering methods were applied in the video processing to calculate the ice concentration as a function of time. The results derived from run no. 5100 are presented in Figures 9.7, 9.8, and 9.9.

The analysis of the test run no. 5100 indicates that the ice concentration reached a limiting value of around 89% at approximately 200 s after the start of the test. This value is only 3% higher than the target value, and it is, therefore, concluded that the ice sheet was prepared well in this test run.

Figure 9.10 shows the variation of the Otsu method's threshold in time for all test runs. The average ice concentrations after reaching the limiting values in all test runs are summarized in Table 9.3.

Reduced ice concentration in the initial part of the test runs (before convergence) is related to the model vessel positioning. It is an unwanted phenomenon, since it reduces the effective length of the ice tank. It is recommended to develop mitigation procedures in the future to help avoid this issue.

In all test runs, it was observed that the ice concentration in the near vicinity of the model was reaching a limiting value of approximately 80% − 89%, irrespective of the

(a) Run no. 5100: Time=816 s.

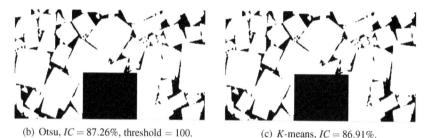

(b) Otsu, $IC = 87.26\%$, threshold $= 100$.   (c) K-means, $IC = 86.91\%$.

Figure 9.7 Frames of run no. 5100 at 816 s and ice detection (Source: Figures from Q. Zhang, S. van der Werff, I. Metrikin, S. Løset and R. Skjetne, " Image Processing for the Analysis of an Evolving Broken-Ice Field in Model Testing," In *ASME 31st International Conference on Ocean, Offshore and Arctic Engineering*, Rio de Janeiro, Brazil, 2012).

starting ice concentrations and floe sizes. This phenomenon can be explained by the tank's wall effect. That is, the ice floes were compacted by the model vessel toward the end of the basin, such that the ice concentration asymptotically approached a limiting value.

The results of the video processing confirm that the difference between the Otsu and the k-means methods is quite small. The differences between the target ice concentration and the actual values obtained from the image processing indicate that the broken ice sheet preparation procedures could be improved. Specifically, more attention could be paid to preparing ice sheets with low ice concentrations ($< 80\%$). By comparing the original images to the processed ones, both of these methods are effective if there is only one type of ice on the water.

9.3 ICE FLOE IDENTIFICATION

The proposed GVF snake-based ice floe identification algorithm is adopted to identify individual model sea ice floes. Ice floe characteristics such as position, area, and size distribution are then obtained. Based on the characters of the model sea ice floes, a model of the managed ice field's configuration, including identification of overlapping floes, is presented here for further studies in ice-force numerical simulations. Finally, the ice floe identification algorithm is applied to an ice surveillance video to further illustrate its applicability to ice management.

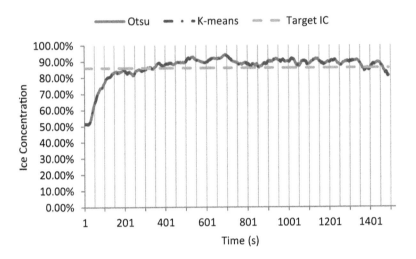

Figure 9.8 Time-varying *IC* of run no. 5100 based on Otsu and *k*-means. Target *IC* = 86% (Source: Figure from Q. Zhang, S. van der Werff, I. Metrikin, S. Løset and R. Skjetne, "Image Processing for the Analysis of an Evolving Broken-Ice Field in Model Testing," In *ASME 31st International Conference on Ocean, Offshore and Arctic Engineering*, Rio de Janeiro, Brazil, 2012).

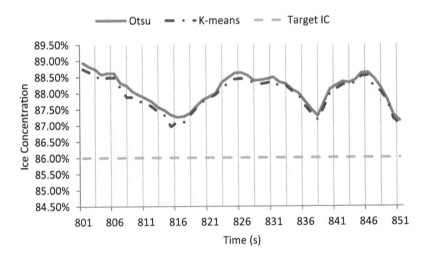

Figure 9.9 Time-varying *IC* of run no. 5100 based on Otsu and *k*-means at 801-851s. Target *IC* = 86% (Source: Figure from Q. Zhang, S.van der Werff, I. Metrikin, S. Løset and R. Skjetne, "Image Processing for the Analysis of an Evolving Broken-Ice Field in Model Testing," In *ASME 31st International Conference on Ocean, Offshore and Arctic Engineering*, Rio de Janeiro, Brazil, 2012).

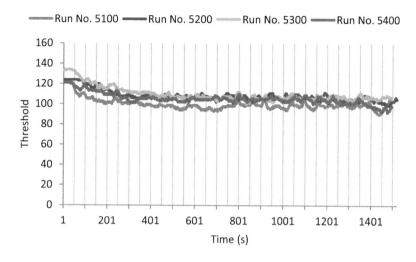

Figure 9.10 Time-varying *IC* of run no. 5100-5400 based on Otsu thresholding (Source: Figure from Q. Zhang, S. van der Werff, I. Metrikin, S. Løset and R. Skjetne, " Image Processing for the Analysis of an Evolving Broken-Ice Field in Model Testing," In *ASME 31st International Conference on Ocean, Offshore and Arctic Engineering*, Rio de Janeiro, Brazil, 2012).

Table 9.3

Average *IC* after reaching saturation in all test runs.

Run no.	5100	5200	5300	5400
Start time (s)	200	300	600	300
Average *IC*	88.93%	80.39%	81.69%	84.83%

Source: Table from Q. Zhang, S. van der Werff, I. Metrikin, S. Løset and R. Skjetne, " Image Processing for the Analysis of an Evolving Broken-Ice Field in Model Testing," In *ASME 31st International Conference on Ocean, Offshore and Arctic Engineering*, Rio de Janeiro, Brazil, 2012.

9.3.1 CONTOUR INITIALIZATION FOR CROWDED RECTANGULAR-SHAPED MODEL SEA ICE FLOES

In the model sea ice test, when multiple square-shaped model sea ice floes crowded together (as shown in Figure 9.11(a)), particularly if the floes were aligned in the image, it can be difficult to find a "hole" between the connected floes after binarizing the image by either the thresholding or the *k*-means clustering method, or the combination of these two methods. Hence, it increases the difficulty of locating the seeds for each ice floe (as shown in Figure 9.11(b)) with the result that some connected ice floes could not be separated (as shown in Figure 9.11(c)). It follows that an additional round of contour initialization and segmentation is necessary.

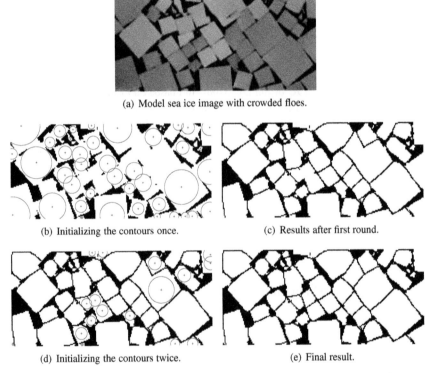

(a) Model sea ice image with crowded floes.

(b) Initializing the contours once.   (c) Results after first round.

(d) Initializing the contours twice.   (e) Final result.

Figure 9.11   Crowded model sea ice floes segmentation (Source: Figures from Q. Zhang, R. Skjetne, I. Metrikin, and S. Løset, "Image Processing for Ice Floe Analyses in Broken-ice Model Testing," *Cold Regions Science and Technology*, 111:27-38, 2015).

Since the ice floes were modeled square shapes with predefined side lengths in this model sea ice test, the largest floe has an area less than a predefined value. Although they are not perfect squares, most of the floes could be approximated as rectangles with a length-to-width ratio less than the given threshold. Based on these characteristics, we use three criteria to determine whether it is necessary to initialize the contours and conduct a second segmentation:

1. The ice floe area is less than the given threshold.
2. The ice floe has a convex shape (the ratio between the floe area and its minimum area-bounding polygon area is larger than the threshold).
3. The length-to-width ratio of the minimum area-bounding rectangle of the ice floe is less than the threshold.

After a segmentation step, the algorithm will stop if all the identified floes satisfy these criteria. Otherwise, the algorithm must find the floes that do not satisfy any of these criteria, calculate their distances, find the new seeds, initialize new contours, and perform the segmentation again. Some boundaries may exist that are too weak to be detected, and there may be some floes that do not satisfy the criteria after a new step. However, the total number of identified floes will converge to a final solution. Therefore, the algorithm is made to stop if the total number of floes identified after steps N and $N + 1$ are equal, in combination with an absolute stop criterion.

Thus, the following algorithm is proposed to segment the (rectangular-shaped) model sea ice floe image. First, the GVF is derived from the grayscale input image. Then, the ice floes are separated from water by using the thresholding method, and the floes are labeled. Each labeled floe should then be checked to determine whether it satisfies all of the criteria mentioned above. Next, the seeds and radii of the floes that do not satisfy any of the criteria are found. Based on the seeds and radii, the circles are initialized and the snake algorithm is run on those circles. The algorithm will stop when it meets the maximum iteration time, or the total number of floes identified after steps N and $N + 1$ are equal. The pseudocode of the proposed algorithm, as well as a maximum iteration time, is given in Algorithm 7.

Note, this model sea ice floe segmentation algorithm works only with the rectangular-shaped objects, which is an extreme case. For the segmentation of the crowded model sea ice floes that have other shapes, the criteria designed for the square-shaped crowded ice floes in this algorithm can be replaced by the corresponding shape criteria, or even just removed from the algorithm since it is easier to form "holes" between such connected floes after binarization and thereby reduce the difficulty of initializing contours for each of them.

9.3.2　ICE FLOE IDENTIFICATION FOR OVERALL TANK IMAGE

In the run no. 5100, a managed ice condition was obtained by cutting a manufactured level ice layer into square pieces with specific dimensions and distributing them over a specific testing area.

As discussed in Chapter 6, a uniform parameter for the GVF field usually cannot represent the overall ice image. Hence, to derive a more precise result, like the local processing we did in the application of the overall sea ice image previously, the overall image is first divided into several smaller, overlapping sub-images (to avoid image border effects), and the proposed model sea ice floe segmentation algorithm (Algorithm 7) is performed locally on each sub-image. The parameter values of the GVF field for each sub-image are listed in Table 9.4. Then, stitching the segmented

---

Algorithm 7 Model sea ice floe segmentation

Input: Model sea ice image

Start algorithm:

  1:   $T \leftarrow$ max iteration time

  2:   $N_0 \leftarrow 0$

  3:   $GVF \leftarrow$ GVF derived from grayscale of input image

  4:   $BW \leftarrow$ binary image

  5:   $FLOE \leftarrow$ labeled ice floes in BW

  6:   $N_1 \leftarrow$ total number of ice floes in BW

  7:   **if** $N_0 \neq N_1 \&\& T \neq 0$ **then**

  8:      $f \leftarrow$ floes in $FLOE$ that do not satisfy any of the criteria

  9:      $k \leftarrow$ number of f

10:      **if** $k \neq 0$ **then**

11:         $S \leftarrow$ Seeds of f found by local maxima of distance transform

12:         **for** each seed $s \in S$ **do**

13:            $r \leftarrow$ local maxima values at s

14:            $c \leftarrow$ initial contours locate at s with its radius r

15:            $B \leftarrow$ boundary detected by performing the snake algorithm on c

16:            $BW \leftarrow BW$ with B superimposed

17:         **end for**

18:      **end if**

19:      $N_0 \leftarrow N_1$

20:      $T \leftarrow T - 1$

21:      go back to 5

22:   **end if**

23:   **return** BW

Output: Segmented image

---

sub-images results in an overall segmented ice floe image. After that, the black spots inside the segmented ice floes caused by noise are filled in the ice shape enhancement step. The final segmentation result and its floe size distribution histogram are shown in Figures 9.12 and 9.13, respectively.

Several lights were installed at the bottom of the ice tank to supply a sufficient brightness for the ice observation. However, these lights are detrimental to the ice image analysis because they can be identified as ice floes due to their brightness, and the light reflection on the ice floe may induce erroneous segmentation. As shown in Figure 9.14, the initial circle meets a strong light reflection when deforming, and some boundaries around the reflection become too weak to be detected. Hence, a part of the circle deformed toward the reflection rather than toward the true floe boundary. Fortunately, the light reflection did not significantly affect our segmentation result, but we still suggest that the lights be disabled or that a polarizer be placed in front of the camera before taking the picture.

Table 9.4

The parameter values of the GVF field for each sub-image.

Sub-image no.	1	2	3	4	5	6	7
The GVF iteration number	150	65	65	65	160	60	110
Sub-image no.	8	9	10	11	12	13	14
The GVF iteration number	130	90	130	170	160	100	90
Sub-image no.	15	16	17	18	19	20	
The GVF iteration number	90	100	90	80	90	80	

Source: Table from Q. Zhang, R. Skjetne, I. Metrikin, and S. Løset, "Image Processing for Ice Floe Analyses in Broken-ice Model Testing," *Cold Regions Science and Technology*, 111:27-38, 2015.

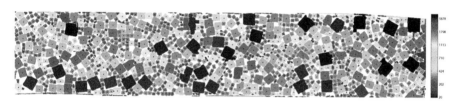

Figure 9.12 Overall model sea ice floe identification (Source: Figure from Q. Zhang, R. Skjetne, I. Metrikin, and S. Løset, "Image Processing for Ice Floe Analyses in Broken-ice Model Testing," *Cold Regions Science and Technology*, 111:27-38, 2015).

9.3.2.1 Model sea ice floe modeling

In the numerical simulation of the ice-structure interaction, all of the ice floes are modeled as rectangular floes, and the positions of the vertices are important to an ice-structure analysis. Therefore, we proposed to perform ice floe rectangularization. Ice floe rectangularization is achieved by assigning the minimum area-bounding rectangle to each ice floe. Due to under- and over-segmentation, the rectangles with a length-to-width ratio less than a given threshold are removed. The final rectangularization result is shown in Figure 9.15(a). If the floes are not segmented well, the rectangularized floes will be smaller or larger than the actual segmented floes. Furthermore, because of the rectangularization, some rectangularized floes will overlap, as seen in Figure 9.15(b). Similar to the sea ice modeling as introduced in Section 8.2.1, besides the vertices, center, area, and perimeter, the "flag" indicating the overlaps between ice floes is also registered to each rectangular floe in our database.

The colorized histogram of the rectangular floe size distribution is presented in Figure 9.16(a). Under-segmentation could induce a large area difference as well as

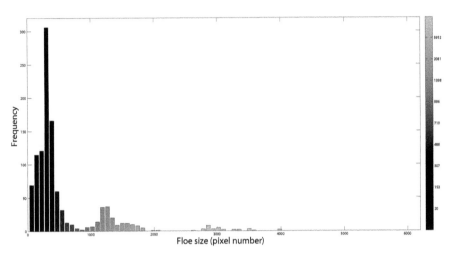

Figure 9.13   Ice floe size distribution histogram of Figure 9.12 (Source: Figure from Q. Zhang, R. Skjetne, I. Metrikin, and S. Løset, "Image Processing for Ice Floe Analyses in Broken-ice Model Testing," *Cold Regions Science and Technology*, 111:27-38, 2015).

overlapping, which explains why the largest floe in Figure 9.15(a) is much larger than the largest identified floe in Figure 9.12, as seen in Figure 9.16(b) by comparing Figures 9.16(a) and 9.13.

Note that, the target ice concentration in this test was 86%. The ice concentration derived from Figure 9.12 is 76.96%, while it is 83.17% when calculated using the threshold method (estimated by counting the number of pixels for each respective area in the image). In the proposed algorithm, the ice pixels detected as a boundary were changed to water pixels, so the calculated ice concentration was reduced in Figure 9.12. The ice concentration calculated by summing all the areas of the rectangular floes in Figure 9.15(a) over the image domain is 87.75%. This value is slightly higher than the thresholding result because the overlapping parts have been identified and considered. The overlapping parts compensate for the loss of ice concentration, and increase the calculated ice concentration closer to the target value.

9.3.3   ICE FLOE IDENTIFICATION FOR MODEL SEA ICE VIDEO: MONITORING MAXIMUM FLOE SIZE

Another application of the proposed algorithm is to monitor the maximum floe size entering the protected vessel from a physical ice management operation. Similar to the preprocessing we did in the calculation of ice concentration, the motion captured by the video of run no. 5100 is retrieved by analyzing at one frame per second and the impediments around the tank are removed as shown in Figure 9.17(a). Then the proposed algorithm is applied to segment the connected ice floe (Figure 9.17(b)), and the maximum floe area for each frame is calculated. Figure 9.18 presents the maximum floe size entering the protected vessel as a function of time. Based on this

Figure 9.14 Light reflection impact, which may induce erroneous segmentation (Source: Figure from Q. Zhang, R. Skjetne, I. Metrikin, and S. Løset, "Image Processing for Ice Floe Analyses in Broken-ice Model Testing," *Cold Regions Science and Technology*, 111:27-38, 2015).

(a) Rectangularization of Figure 9.12.

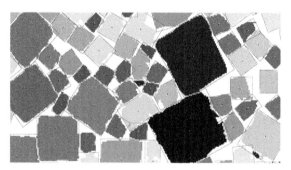

(b) Comparison between identification and rectangularization results. The rectangles are the boundaries of the modified floes. The black '·' are the centers of identified ice floes, and the white '·' are the centers of rectangularized floes.

Figure 9.15 Ice floe rectangularization (Source of Figure 9.15(a): Figure from Q. Zhang, R. Skjetne, I. Metrikin, and S. Løset, "Image Processing for Ice Floe Analyses in Broken-ice Model Testing," *Cold Regions Science and Technology*, 111:27-38, 2015).

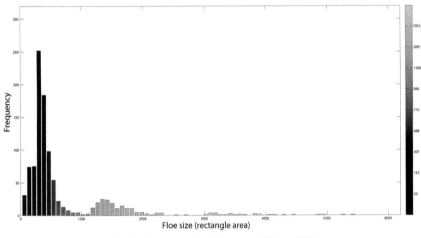

(a) Ice floe size distribution histogram of Figure 9.15(a).

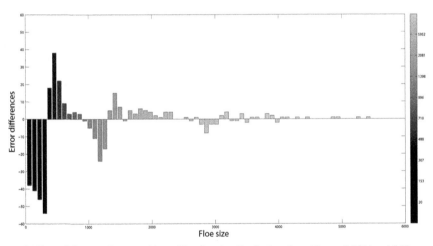

(b) Error differences between the resulting floe size distributions from Figures 9.16(a) and 9.13.

Figure 9.16 Rectangular floe size distribution and the error in the simplified distribution (Source of Figure 9.16(a): Figure from Q. Zhang, R. Skjetne, I. Metrikin, and S. Løset, "Image Processing for Ice Floe Analyses in Broken-ice Model Testing," *Cold Regions Science and Technology*, 111:27-38, 2015).

result, a warning can be sent to the risk management system if the estimated risk based on the maximum floe size is too large.

(a) Pre-processed frame at 965 s.   (b) Segmentation result of Figure 9.17(a).

Figure 9.17 Model sea ice video processing (Source: Figure from Q. Zhang, R. Skjetne, I. Metrikin, and S. Løset, "Image Processing for Ice Floe Analyses in Broken-ice Model Testing," *Cold Regions Science and Technology*, 111:27-38, 2015).

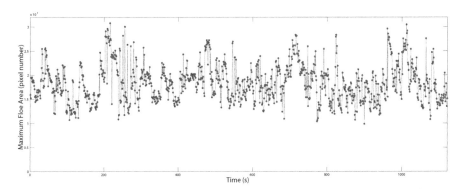

Figure 9.18 Maximum floe size entering the protected vessel (Source: Figure from Q. Zhang, R. Skjetne, I. Metrikin, and S. Løset, "Image Processing for Ice Floe Analyses in Broken-ice Model Testing," *Cold Regions Science and Technology*, 111:27-38, 2015).

A Geometric Calibration

A.1 ORTHORECTIFICATION

The orthorectification methods introduced in this section are based on the ideal pin-hole camera model, in which the lens distortion of the camera is ignored, as illustrated in Figure A.1. The center of the camera is known as the optical center. The line perpendicular to the focal plane through the optical center is the optical axis (Z-axis shown in Figure A.1). The projection of a point P in a 3-dimensional scene with the coordinates (X,Y,Z) to the corresponding point p in the image is a mapping from Euclidean 3-dimensional space to Euclidean 2-dimensional space, given by [16]:

$$x = \frac{-fX}{Z-f} \tag{A.1a}$$

$$y = \frac{-fY}{Z-f} \tag{A.1b}$$

where x and y are the image coordinates of p, and f is the focal length of the camera lens.

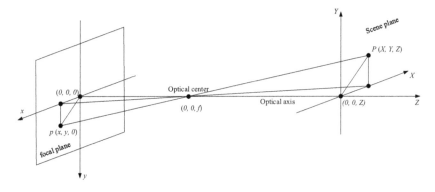

Figure A.1 Pinhole camera imaging system.

A.1.1 AN ANALYTICAL METHOD

The location of any planar object in the image is a function of the spatial orientation of the camera in relation to ground topography. Figure A.2 illustrates the relationship between the image coordinates and the orthorectification coordinates. Note, in this figure, the focal plane is placed in front of the optical center of the camera.

The image coordinates lay in the focal plane and are denoted with small letters (x,y). The orthorectification plane coordinates are parallel to the ground and are denoted with capital letters (X,Y). The optical center of the camera is S. The camera nadir line intersects the orthorectification plane at the nadir T. The optical axis is perpendicular to the focal plane and intersects the center of the focal plane at principal point r, forming the shooting angle φ with the vertical nadir line. The optical axis extends to the orthorectification plane at point R. The principal line passes through point R (r) and the image border at O (or o), and it bisects the orthorectification and focal planes. The point O (or o) acts as the origin for the image coordinate system with the y-axis as the principal line, and the origin for the orthorectification coordinate system with the principal line in the orthorectification plane defining the positive Y-axis. $p(x,y)$ is any point on the focal plane, while $P(X,Y)$ is the corresponding point on the orthorectification plane.

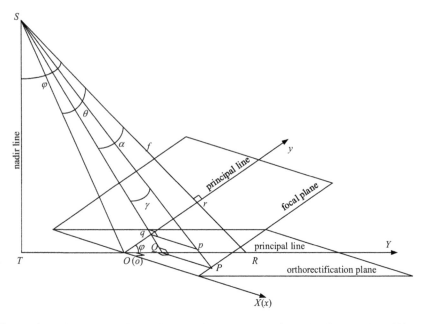

Figure A.2 Geometric orthorectification (Source: Figure from Q. Zhang and R. Skjetne, "Image Processing for Identification of Sea-Ice Floes and the Floe Size Distributions," *IEEE Transactions on Geoscience and Remote Sensing*, 53(5):2913-2924, 2015).

From Figure A.2, we obtain:

$$TO = SO\sin(\varphi - \theta) = f\sec\theta\sin(\varphi - \theta) \tag{A.2}$$

$$TQ = ST\tan(\varphi + \alpha) = f\sec\theta\cos(\varphi - \theta)\tan(\varphi + \alpha) \tag{A.3}$$

$$SQ = ST\sec(\varphi + \alpha) = f\sec\theta\cos(\varphi - \theta)\sec(\varphi + \alpha) \tag{A.4}$$

$$\gamma = \arctan(qp/Sq) = \arctan\frac{qp\cos\alpha}{f}, \tag{A.5}$$

where θ is half of the lens' vertical field of view angle, f is the focal length, and $\alpha = \arctan(qr/f)$. It should be aware of the direction of qr, and the signs of α and γ.

According to Equations (A.2)-(A.5), we derive

$$OQ = TQ - TO$$

$$= f \sec\theta \left[\cos(\varphi - \theta) \tan\left(\varphi + \arctan\frac{qr}{f} \right) - \sin(\varphi - \theta) \right] \tag{A.6}$$

$$QP = SQ\tan\gamma = qp\frac{\cos\alpha\cos(\varphi - \theta)}{\cos\theta\cos(\varphi + \alpha)}$$

$$= qp\frac{f}{\sqrt{f^2 + qr^2}}\frac{\cos(\varphi - \theta)}{\cos\theta\cos\left(\varphi + \arctan\frac{qr}{f} \right)} \tag{A.7}$$

A digital image is a numeric representation of a 2-dimensional picture. It is composed of pixels that are the smallest individual elements of the image. We assume that the pixel magnification of the image is μ, yielding

$$\begin{cases} oq = y \cdot \mu \\ qp = x \cdot \mu \end{cases} \tag{A.8}$$

$$\begin{cases} OQ = Y \cdot \mu \\ QP = X \cdot \mu \end{cases} \tag{A.9}$$

Therefore,

$$qr = oq - or = \left(y - \frac{n_y}{2} \right) \cdot \mu \tag{A.10}$$

where n_y is the number of pixels in the image length. It should also be aware of the directions of qp, QP, and qr.

Instead of working from a $1:1$ positive, we counted image pixels on a computer screen. Hence, the apparent focal length of the image is altered by

$$f' = \frac{y_e}{\tan\theta} = \frac{\mu \cdot n_y/2}{\tan\theta}, \tag{A.11}$$

where y_e is the half-length of the image. Substituting Equations A.8 to A.11 into Equations A.6 and A.7, the location of any point $P(X,Y)$ in the orthorectification coordinates can be determined from its image coordinates $p(x,y)$ by:

$$\begin{cases} Y = \dfrac{n_y}{2} \csc\theta \left\{ \cos(\varphi - \theta)\tan\left[\varphi + \arctan\left(\dfrac{y - n_y/2}{n_y/2}\tan\theta \right) \right] - \sin(\varphi - \theta) \right\} \\ X = x \cdot \dfrac{\frac{n_y}{2}\csc\theta\cos(\varphi - \theta)\sec\left[\varphi + \arctan\left(\frac{y - n_y/2}{n_y/2}\tan\theta \right) \right]}{\sqrt{\left(\frac{n_y/2}{\tan\theta} \right)^2 + \left(y - \frac{n_y}{2} \right)^2}} \end{cases}$$

$$\tag{A.12}$$

This ends the orthorectification procedure.

From Equation A.12, we find that, although μ varies with the magnification ratio of the image, the relationship between different coordinates will not change. This is because the influence of μ is counteracted in Equations A.8 to A.11. Consequently, the relationship between the image coordinates and the orthorectification coordinates is a function of the shooting angle φ and the camera's (vertical) field of view 2θ.

An example of the geometric orthorectification by using the analytical method is shown in Figure A.3

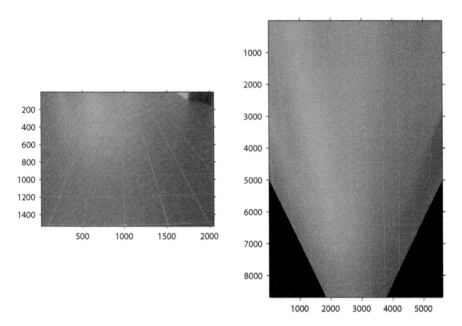

Figure A.3 An example of the geometric orthorectification by using the analytical method. Left: distorted image. Right: calibrated image.

A.1.2 A LINEAR APPROXIMATION

The images of any planar object observed from different views can be related by a direct linear transformation (DLT) in the projective space, given by [3]:

$$\mathbf{x}' = \mathbf{H}\mathbf{x} \tag{A.13}$$

where $\mathbf{x} = [x, y, 1]^{\mathrm{T}}$ and $\mathbf{x}' = [x', y', 1]^{\mathrm{T}}$ are 3-vectors corresponding to the image planes of the same point, denoted by $p(x, y)$ and $p'(x', y')$, respectively, and

$$\mathbf{H} = \begin{bmatrix} h_1 & h_2 & h_3 \\ h_4 & h_5 & h_6 \\ h_7 & h_8 & 1 \end{bmatrix} \tag{A.14}$$

is a 3×3 matrix consisting of 8 degrees of freedom (numeral 1 in H is the scale invariance). This linear transformation is referred to as a homography that maps any point $p(x, y)$ on image plane π to the corresponding point $p'(x', y')$ on image plan π'.

From Equation A.13, x' and y' can be solved by:

$$x' = \frac{h_1 x + h_2 y + h_3}{h_7 x + h_8 y + 1} \tag{A.15a}$$

$$y' = \frac{h_4 x + h_5 y + h_6}{h_7 x + h_8 y + 1} \tag{A.15b}$$

Equations A.15a and A.15b can be re-written as:

$$h_1 x + h_2 y + h_3 - h_7 x x' - h_8 y x' = x' \tag{A.16a}$$
$$h_4 x + h_5 y + h_6 - h_7 x y' - h_8 y y' = y' \tag{A.16b}$$

These two equations can be expressed in matrix form as:

$$\begin{bmatrix} x & y & 1 & 0 & 0 & 0 & -xx' & -yx' \\ 0 & 0 & 0 & x & y & 1 & -xy' & -yy' \end{bmatrix} \mathbf{h} = \begin{bmatrix} x' \\ y' \end{bmatrix} \tag{A.17}$$

where $\mathbf{h} = [h_1, h_2, h_3, h_4, h_5, h_6, h_7, h_8]^T$.

The transformation matrix H has 8 unknown coefficients, which means that at least 8 variables are required to determine H. Each point pair provides two equations. With known correspondences between at least 4 non-aligned points, respectively denoted by (x_i, y_i) and (x'_i, y'_i), $i = 1, 2, 3, 4$, the coefficients in H can thereby be estimated by solving the matrix equation:

$$\mathbf{Ah} = \mathbf{b} \tag{A.18}$$

where

$$\mathbf{A} = \begin{bmatrix} x_1 & y_1 & 1 & 0 & 0 & 0 & -x_1 x'_1 & -y_1 x'_1 \\ 0 & 0 & 0 & x_1 & y_1 & 1 & -x_1 y'_1 & -y_1 y'_1 \\ x_2 & y_2 & 1 & 0 & 0 & 0 & -x_2 x'_2 & -y_2 x'_2 \\ 0 & 0 & 0 & x_2 & y_2 & 1 & -x_2 y'_2 & -y_2 y'_2 \\ x_3 & y_3 & 1 & 0 & 0 & 0 & -x_3 x'_3 & -y_3 x'_3 \\ 0 & 0 & 0 & x_3 & y_3 & 1 & -x_3 y'_3 & -y_3 y'_3 \\ x_4 & y_4 & 1 & 0 & 0 & 0 & -x_4 x'_4 & -y_4 x'_4 \\ 0 & 0 & 0 & x_4 & y_4 & 1 & -x_4 y'_4 & -y_4 y'_4 \end{bmatrix}$$

and $\mathbf{b} = [x'_1, y'_1, x'_2, y'_2, x'_3, y'_3, x'_4, y'_4]^T$ for the minimum requirement.

The solution for $\mathbf{h}$ can be found by using standard techniques for solving linear equations, such as Gaussian elimination. In the case of an over-determined set of equations where $\mathbf{A}$ contains more than 8 rows (i.e., more than 4 correspondences are used for generating the matrix Equation A.18), the least-squares technique can be adopted for estimating $\mathbf{h}$ [57].

A.2   CALIBRATION OF RADIAL LENS DISTORTION

A.2.1   AN ANALYTICAL METHOD FOR FISHEYE DISTORTION

The location of any object in the image is a function of the field of view (FOV) angle of the camera in relation to its calibrated flat surface. Figure A.4 illustrates the mapping between fisheye image coordinates and the corresponding calibration coordinates.

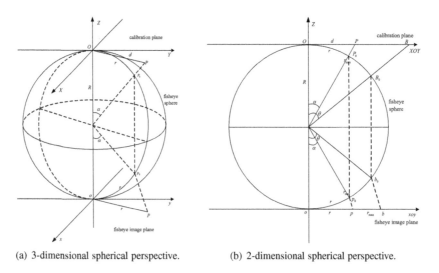

(a) 3-dimensional spherical perspective.   (b) 2-dimensional spherical perspective.

Figure A.4   Fisheye calibration (Source: Figure from Q. Zhang and R. Skjetne, "Image Techniques for Identifying Sea-Ice Parameters," *Modeling, Identification and Control*, 35(4):293-301, 2014).

The fisheye image coordinates, denoted with small letters (x, y), is located at the bottom of the fisheye sphere and is perpendicular to the Z-axis, which is the optical axis of the lens. The calibration coordinates, denoted with capital letters (X, Y), are on the top of the fisheye sphere parallel to the image coordinates. The center of the fisheye image o located in the Z-axis, acts as the origin of the image coordinates. The intersection point O between the calibration plane and the Z-axis acts as the origin of the calibration coordinates. θ is half known FOV angle of the lens. For any point $p(x, y)$ on the image plane and its corresponding point $P(X, Y)$ on the calibration plane, the radial distance $|OP|$ in the distorted fisheye image plane $|op|$ is equivalent to the length of the arc segment r between the Z-axis and the point P_0, which is the intersection point of the projection ray of the point P and the fisheye sphere. Therefore,

$$|op| = r = R \cdot \alpha = R \cdot \arctan \frac{|OP|}{R} \tag{A.19}$$

where R is the radius of the fisheye sphere. This is obtained by:

$$R = \frac{r_{max}}{\theta} = \frac{r_m \cdot \mu}{\theta} \tag{A.20}$$

where r_{max} is half the capture range of the fisheye image in length, r_m is the pixel number in maximum semidiameter, obtained by counting half the number of pixels in the image length, and μ is the pixel magnification of the image. From Figure A.4(a), we also find that

$$|op| = \sqrt{x^2 + y^2} \cdot \mu \tag{A.21a}$$

$$|OP| = \sqrt{X^2 + Y^2} \cdot \mu \tag{A.21b}$$

$$\frac{x}{y} = \frac{X}{Y} \tag{A.21c}$$

Substituting Equations A.20 to A.21c into A.19, the location of any point $P(X,Y)$ in the calibration coordinates can then be determined from its fisheye image coordinates $p(x,y)$ by:

$$X = \frac{x}{\sqrt{x^2 + y^2}} \cdot \frac{r_m \tan \frac{\theta \sqrt{x^2 + y^2}}{r_m}}{\theta} \tag{A.22a}$$

$$Y = \frac{y}{\sqrt{x^2 + y^2}} \cdot \frac{r_m \tan \frac{\theta \sqrt{x^2 + y^2}}{r_m}}{\theta} \tag{A.22b}$$

From Equations A.22a and A.22b, we find that the relationship between the fisheye image coordinates and its calibration coordinates only depends on the camera's FOV angle 2θ. The magnification μ will not affect the relationship since it is counteracted in Equations A.20 to A.21b.

An example of the fisheye calibration by using the analytical method is shown in Figure A.5

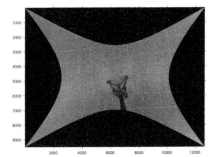

Figure A.5 An example of the fisheye calibration by using the analytical method. Left: distorted image. Right: calibrated image.

A.2.2   A POLYNOMIAL APPROXIMATION

The radial distortion of a perfectly centered lens can be modeled as [170]:

$$\begin{bmatrix} x_d \\ y_d \end{bmatrix} = L(r) \begin{bmatrix} x \\ y \end{bmatrix} \tag{A.23}$$

where (x_d, y_d) is the position of a point in the distorted image and (x, y) is the corresponding position in the ideal undistorted image after calibration (note here that (x, y) follows linear projection [57]), $r = \sqrt{x^2 + y^2}$ is the radial distance from the center of the distorted image, and

$$L(r) = 1 + \kappa_1 r^2 + \kappa_2 r^4 + \cdots \tag{A.24}$$

is the distortion factor with the coefficients $\kappa_1, \kappa_2, \cdots$. The distortion coefficient κ_1 in Equation A.24 could dominate the other coefficients and achieve an accuracy of about 0.1 pixels in the image space by using lenses exhibiting large distortion [160]. The radial distortion is called a barrel distortion when $\kappa_1 < 0$, while it is called a pincushion distortion for $\kappa_1 > 0$, as seen in Figure A.6.

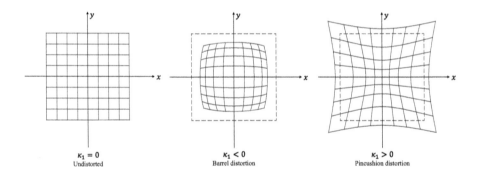

Figure A.6   Radial distortion.

The function $L(r)$ can be determined by using the deviation of several known reference points from their original positions in a distorted image. Another approach is based on the fact that the straight lines in the scene should appear straight in the image [41]. Thus $L(r)$ can be estimated by mapping these radial distorted lines to straight lines.

B Ice Image Data Structure

All the processed ice image parameters and data are stored in this MATLAB® ice image data structure. At data capture, the following parameters are defined and stored:

IceImage Ice image data structure.

    .Param   Global image parameters all stored with image at image capture.

        .NumPix_x Number of pixels in x-direction.

        .NumPix_y Number of pixels in y-direction.

        .MAMSL Meters Above Mean Sea Level of camera [meter].

        .FOV   Field of View angle [degrees].

        .TiltAngle Camera lens tilt angle (vertical shooting angle) [degrees].

        .PanAngle Camera lens panning angle (horizontal shooting angle) [degrees].

        .CamRot Rotation angles of camera w.r.t. vessel sensor platform coordinate system [degrees].

        .LeverArm Lever arm $[x, y, z]$ of camera w.r.t. vessel sensor platform coordinate system [meter].

        .VesselRot Orientation angles [Roll, Pitch, Yaw] of vessel [degrees].

        .VesselPos Geographical position of vessel coordinate system [Lat, Long].

        .FileName Image file name [string].

        .Time   Exact date and time of image capture [string].

        .Location Descriptive location of image capture [string].

        .Creator Photographer or creator of image [string].

        .Caption Image caption text (string) [string].

        .PrjName Project name related to ice image acquisition [string].

        .PrjNum Project number related to ice image acquisition [string].

    .Field   Data structures with global information about the ice field.

        .Georef Georeference of origin of image coordinate system [Lat, Long].

        .Rot   Rotation angle of ice field x-axis w.r.t. North [degrees].

        .LengthSI_x Length of ice field domain in meters in x-direction [meter].

        .LengthSI_y Length of ice field domain in meters in y-direction [meter].

        .PixScale_x Length scale of each pixel in x-direction at sea surface [meter/pixel].

.PixScale_y Length scale of each pixel in y-direction at sea surface [meter/pixel].

.PixArea Area of each pixel at sea surface [meter^2].

.NumFloes Number of detected ice floes.

.NumBrash Number of detected brash ice pieces.

.CovFloes Coverage of ice floe pixels in the image.

.CovBrash Coverage of brash ice pixels in the image.

.CovSlush Coverage of slush ice pixels in the image.

.CovWater Coverage of open water pixels in the image.

.CovOther Coverage of other residual pixels in the image.

.FSD Floe-size distribution given by cell array $\{[\text{int\_min}, \text{int\_max}, \text{num}], \cdots\}$ where each defined interval is given by a triplet of floe area interval minimum, interval maximum, and number of floes.

.Floe(i) Array of data structures that each contain information about a single ice floe.

.Center Pixel coordinates of center point of floe [index_x, index_y].

.Area Area of floe in number of pixels.

.Perimeter Perimeter of each floe in unit of pixels.

.Polygon Substructure with data corresponding to polygon (convex hull) fit to the true ice floe:

.Vertices 2D array of pixel coordinates of vertices corresponding to a convex hull fit of each floe. There shall be no duplicated vertices, assuming the last vertex connects to the first vertex.

.Center Pixel coordinates of center point of polygon.

.Area Area of polygon in number of polygon.

.Perimeter Perimeter of polygon in unit of pixels.

.Intersect Substructure corresponding to intersection between ice floes:

.floe Array of indices of other floes that current floe intersects with, e.g.: Intersect.floe = $[6, 13, 34]$.

.brash Array of indices of brash pieces that current floe intersects with, e.g.: Intersect.brash = $[57, 60]$.

.Pixels Cell array of indices of all pixels making out the individual real identified floe in the overall image.

.Brash(i) Array of data structures that each contain information about a single brash ice piece.

.Center Pixel coordinates of center point of brash piece.

.Area Area of brash piece in number of pixels.

.Circle Substructure with data corresponding to circular fit to the true brash piece.

.Radius   Radius of brash circle in unit of pixels.

.Perimeter Perimeter length of circle in unit of pixels.

.Intersect Substructure corresponding to intersection data:

  .floe Array of indices of floes that current brash piece intersects with, e.g.: Intersect.floe = $[6, 13, 34]$.

  .brash Array of indices of other brash pieces that current brash piece intersects with, e.g.: Intersect.brash = $[57, 60]$.

.Pixels Cell array of indices of all pixels making out the individual real identified brash piece in the overall image.

Figure B.1 gives an example of an ice image data structure derived from Figure 8.8, and Figures B.2 and B.3 show the examples of the parameters we stored for each ice floe and brash ice in this ice image data structure.

IceImage ✕	
⊞ 1x1 struct with 4 fields	
Field ▲	**Value**
⊞ Param	1x1 struct
⊞ Field	1x1 struct
⊞ Floe	498x1 struct
⊞ Brash	201x1 struct

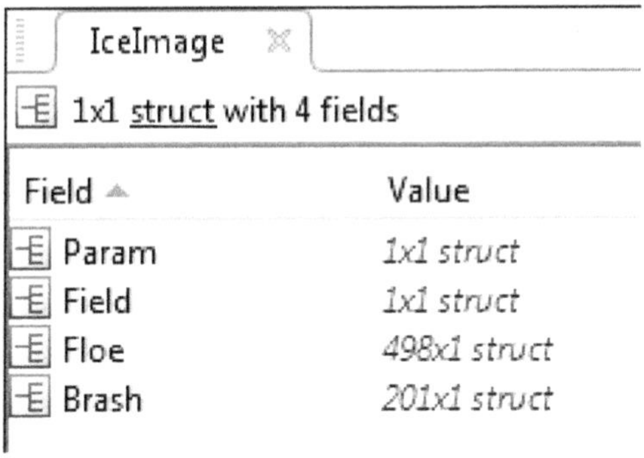

Figure B.1 Ice image data structure.

	IceImage ✕	IceImage.Floe ✕		

📇 498x1 struct with 5 fields

Fields	🗂 Center	⊞ Area	⊞ Perimeter	📇 Polygon	🗂 Pixels
1	[84.9048 27...	21	14.4853	1x1 struct	21x2 double
2	[953.3636 1...	22	16.7279	1x1 struct	22x2 double
3	[958.0455 3...	22	15.8995	1x1 struct	22x2 double
4	[152.3810 1...	21	14.4853	1x1 struct	21x2 double
5	[507.7143 3...	21	14.7279	1x1 struct	21x2 double
6	[315.7143 3...	21	14.4853	1x1 struct	21x2 double
7	[889.6667 3...	21	19.5563	1x1 struct	21x2 double
8	[1022 2.4348]	23	16.4853	1x1 struct	23x2 double
9	[583.8095 2....	21	17.0711	1x1 struct	21x2 double
10	[811.9048 2...	21	14.4853	1x1 struct	21x2 double
11	[291.7619 2...	21	22.3848	1x1 struct	21x2 double
12	[1.0013e+03...	23	15.8995	1x1 struct	23x2 double
13	[121.9565 1...	23	15.8995	1x1 struct	23x2 double
14	[228.0455 3...	22	15.8995	1x1 struct	22x2 double
15	[249.8696 7...	23	15.8995	1x1 struct	23x2 double
16	[419.2500 3...	24	15.8995	1x1 struct	24x2 double

(a) Array of data structures that each contain information about each single ice floe (IceImage.Floe)

	IceImage ✕	IceImage.Floe ✕	IceImage.Floe(12).Pixels ✕

⊞ 23x2 double

	1	2	3	4	5
1	998	140			
2	999	139			
3	999	140			
4	999	141			
5	1000	139			
6	1000	140			
7	1000	141			
8	1000	142			
9	1001	139			
10	1001	140			
11	1001	141			
12	1001	142			
13	1002	138			
14	1002	139			
15	1002	140			
16	1002	141			

	IceImage ✕	IceImage.Floe ✕	IceImage.Floe(12).Polygon ✕

📇 1x1 struct with 5 fields

Field ⌃	Value	Min	Max
⊞ Vertices	8x2 double	138	1004
⊞ Center	[1.0012e+03 139.9899]	139.98...	1.0012...
⊞ Area	16.5000	16.5000	16.5000
⊞ Perimeter	15.6476	15.6476	15.6476
📇 Intersect	1x1 struct		

(b) Cell array of indices of all pixels making out the individual real identified floe No.12 in the overall image (IceImage.Floe(12).Pixels).

(c) Substructure with data corresponding to polygon (convex hull) fit to the true ice floe No.12 (IceImage.Floe(12).Polygon).

Figure B.2 Database of ice floes.

IceImage ✕ IceImage.Brash ✕			
⊟ 201x1 struct with 4 fields			

Fields	⊡ Center	⊞ Area	⊟ Circle	⊡ Pixels
123	[281 260]	5	1x1 struct	[280 260;28...
124	[305.8333 2...	6	1x1 struct	6x2 double
125	[741 246]	5	1x1 struct	[740 246;74...
126	[778.6000 1...	5	1x1 struct	[778 178;77...
127	[966.5000 95]	8	1x1 struct	8x2 double
128	[38.8667 10...	15	1x1 struct	15x2 double
129	[43.7000 74....	10	1x1 struct	10x2 double
130	[157.3571 1...	14	1x1 struct	14x2 double
131	[158.6000 1...	15	1x1 struct	15x2 double
132	[900 26]	5	1x1 struct	[899 26;900 ...
133	[928.2143 2...	14	1x1 struct	14x2 double
134	[967.2308 1...	13	1x1 struct	13x2 double
135	[3.2143 345....	14	1x1 struct	14x2 double
136	[126 61.5000]	8	1x1 struct	8x2 double
137	[663 283.50...	8	1x1 struct	8x2 double
128	[605 105 50	12	1x1 struct	12x2 double

(a) Array of data structures that each contain information about each single brash ice piece (IceImage.Brash)

IceImage ✕ IceImage.Brash ✕ IceImage.Brash(131).Pixels ✕				
⊞ 15x2 double				

	1	2	3	4	5
1	157	138			
2	157	139			
3	158	137			
4	158	138			
5	158	139			
6	158	140			
7	158	141			
8	159	138			
9	159	139			
10	159	140			
11	159	141			
12	159	142			
13	160	139			
14	160	140			
15	160	141			
16					

IceImage ✕ IceImage.Brash ✕ IceImage.Brash(131).Circle ✕			
⊟ 1x1 struct with 3 fields			

Field ⌃	Value	Min	Max
⊞ Radius	2.1851	2.1851	2.1851
⊞ Perimeter	13.7294	13.7294	13.7294
⊟ Intersect	1x1 struct		

(b) Cell array of indices of all pixels making out the individual real identified brash piece No.131 in the overall image (IceImage.Brash(131).Pixels).

(c) Substructure with data corresponding to circular fit to the true brash piece No.131 (IceImage.Brash(131).Circle).

Figure B.3 Database of brash ice pieces.

Glossary

CFL: Courant-Friedrichs-Lewy.
CIS: Canadian Ice Service.
CMY: cyan, magenta, and yellow.
DEM: discrete element method.
DP: dynamic positioning.
ERS: European remote sensing.
EP: erosion-propagation.
FOV: field of view.
FSD: floe size distribution.
GLCM: gray-level co-occurrence matrices.
GVF: gradient vector flow.
HSI: hue, saturation, and intensity.
HSVA: Hamburg Ship Model Basin.
IB: ice breaker.
IC: ice concentration.
IMU: inertial measurement unit.
IRGS: iterative region growing using semantics.
LoG: Laplacian of Gaussian.
LR: likelihood ratio.
LVQ: learning vector quantization.
MCD: mean clipper diameter.
MIZ: marginal ice zone.
ML: maximum likelihood.
MRF: Markov random field.
NTNU: Norwegian University of Science and Technology.
OATRC: Oden Arctic Technology Research Cruise.
RGB: red, green, and blue.
SAR: synthetic aperture radar.
SAMCoT: Sustainable Arctic Marine and Coastal Technology.
SIMBA: sea ice mass balance in Antarctic.
SPRS: Swedish polar research secretariat.
SSM/I: special sensor microwave/imager.
UAV: unmanned aerial vehicle.

References

1. All about sea ice: Environment: Climate. `https://nsidc.org/cryosphere/seaice/environment/global_climate.html`.
2. All about sea ice: Introduction. `https://nsidc.org/cryosphere/seaice/index.html`.
3. Y. Abdel-Aziz and H.M. Karara. Direct linear transformation into object space coordinates in close-range photogrammetry. In *Proceeding of the Symposium on Close-Range Photogrammetry*, pages 1–18, Urbana-Champaign, 1971. American Society of Photogrammetry.
4. S.T Acton and N. Ray. Biomedical image analysis: tracking. *Synthesis Lectures on Image, Video, and Multimedia Processing*, 2(1):1–152, 2006.
5. P.S.U. Adiga and B.B. Chaudhuri. An efficient method based on watershed and rule-based merging for segmentation of 3-D histo-pathological images. *Pattern Recognition*, 34(7):1449–1458, 2001.
6. J. Banfield. Automated tracking of ice floes: a stochastic approach. *IEEE Transactions on Geoscience and Remote Sensing*, 29(6):905–911, 1991.
7. J.D. Banfield and A.E. Raftery. Ice floe identification in satellite images using mathematical morphology and clustering about principal curves. *Journal of the American Statistical Association*, 87(417):7–16, 1992.
8. S.C. Basak, V.R. Magnuson, G.J. Niemi, and R.R. Regal. Determining structural similarity of chemicals using graph-theoretic indices. *Discrete Applied Mathematics*, 19(1):17–44, 1988.
9. S. Beucher. The watershed transformation applied to image segmentation. *Signal and Image Processing in Microscopy and Microanalysis*, pages 299–314, 1992.
10. S. Beucher and M. Bilodeau. Road segmentation and obstacle detection by a fast watershed transformation. In *Proceedings of IEEE Intelligent Vehicles Symposium*, pages 296–301, Paris, France, 1994. IEEE.
11. S. Beucher and F. Meyer. The morphological approach to segmentation: the watershed transformation. In E. Dougherty, editor, *Mathematical Morphology in Image Processing*, chapter 12, pages 433–481. Marcel Dekker, New York, 1992.
12. J. Beyerer, F.P. León, and C. Frese. *Machine Vision: Automated Visual Inspection: Theory, Practice and Applications*. Springer, 2015.
13. J.D. Blunt, V.Y. Garas, D.G. Matskevitch, J.M. Hamilton, and K. Kumaran. Image analysis techniques for high Arctic deepwater operation support. In *Proceedings of OTC Arctic Technology Conference*, Houston, Texas, USA, 2012. Offshore Technology Conference.
14. A.V. Bogdanov, S. Sandven, O.M. Johannessen, V.Y. Alexandrov, and L.P. Bobylev. Multisensor approach to automated classification of sea ice image data. *IEEE Transactions on Geoscience and Remote Sensing*, 43(7):1648–1664, 2005.
15. J.S. Brown, E.H. Martin, and A. Keinonen. Ice management numerical modeling and modern data sources. In *International Conference and Exhibition on Performance of Ships and Structures in Ice*, Banff, Alberta, Canada, 2012. Society of Naval Architects and Marine Engineers.
16. L.G. Brown. A survey of image registration techniques. *ACM Computing Surveys*

(*CSUR*), 24(4):325–376, 1992.

17. B.A. Burns, M. Schmidt-Grottrup, and T. Viehoff. Methods for digital analysis of AVHRR sea ice images. *IEEE Transactions on Geoscience and Remote Sensing*, 30(3):589–602, 1992.

18. F. Carsey. Review and status of remote sensing of sea ice. *IEEE Journal of Oceanic Engineering*, 14(2):127–138, 1989.

19. V. Caselles, F. Catté, T. Coll, and F. Dibos. A geometric model for active contours in image processing. *Numerische Mathematik*, 66(1):1–31, 1993.

20. V. Caselles, R. Kimmel, and G. Sapiro. Geodesic active contours. *International Journal of Computer Vision*, 22(1):61–79, 1997.

21. T.F. Chan and L.A. Vese. Active contours without edges. *IEEE Transactions on Image Processing*, 10(2):266–277, 2001.

22. Q. Chen, X. Yang, and E.M. Petriu. Watershed segmentation for binary images with different distance transforms. In *Proceedings of the 3rd IEEE International Workshop on Haptic, Audio and Visual Environments and Their Applications*, pages 111–116. IEEE, 2004.

23. X. Chen, X. Zhou, and S.T. Wong. Automated segmentation, classification, and tracking of cancer cell nuclei in time-lapse microscopy. *IEEE Transactions on Biomedical Engineering*, 53(4):762–766, 2006.

24. J. Cheng and J.C. Rajapakse. Segmentation of clustered nuclei with shape markers and marking function. *IEEE Transactions on Biomedical Engineering*, 56(3):741–748, 2009.

25. D.A. Clausi. Comparison and fusion of co-occurrence, Gabor and MRF texture features for classification of SAR sea-ice imagery. *Atmosphere-Ocean*, 39(3):183–194, 2001.

26. D.A. Clausi and H. Deng. Operational segmentation and classification of SAR sea ice imagery. In *IEEE Workshop on Advances in Techniques for Analysis of Remotely Sensed Data*, pages 268–275, Maryland, USA, 2003. IEEE.

27. D.A. Clausi and B. Yue. Comparing cooccurrence probabilities and markov random fields for texture analysis of SAR sea ice imagery. *IEEE Transactions on Geoscience and Remote Sensing*, 42(1):215–228, 2004.

28. F. Cloppet and A. Boucher. Segmentation of overlapping/aggregating nuclei cells in biological images. In *19th International Conference on Pattern Recognition*, pages 1–4. IEEE, 2008.

29. L.D. Cohen. On active contour models and balloons. *CVGIP: Image Understanding*, 53(2):211–218, 1991.

30. L.D. Cohen and I. Cohen. Finite-element methods for active contour models and balloons for 2-D and 3-D images. *IEEE Transactions on Pattern Analysis and Machine Intelligence*, 15(11):1131–1147, 1993.

31. M.J. Collins, C.E. Livingstone, and R.K. Raney. Discrimination of sea ice in the Labrador marginal ice zone from synthetic aperture radar image texture. *International Journal of Remote Sensing*, 18(3):535–571, 1997.

32. G. Comfort, S. Singh, and D. Spencer. Evaluation of ice model test data for moored structures. Technical report, PERD/CHC, 1999.

33. J.C. Comiso. Unsupervised classification of Arctic sea ice using neural network. In *IEEE International Geoscience and Remote Sensing Symposium*, volume 1, pages 414–418. IEEE, 1995.

34. J.C. Comiso and K. Steffen. Studies of Antarctic sea ice concentrations from satellite data and their applications. *Journal of Geophysical Research: Oceans*,

106(C12):31361–31385, 2001.

35. R. Courant, K. Friedrichs, and H. Lewy. On the partial difference equations of mathematical physics. *IBM Journal of Research and Development*, 11(2):215–234, 1967.

36. M.G. Coutinho. *Guide to Dynamic Simulations of Rigid Bodies and Particle Systems*. Springer, 2012.

37. M. Dabboor and M. Shokr. A new likelihood ratio for supervised classification of fully polarimetric SAR data: an application for sea ice type mapping. *ISPRS Journal of Photogrammetry and Remote Sensing*, 84:1–11, 2013.

38. C. Daley, S. Alawneh, D. Peters, B. Quinton, and B. Colbourne. GPU modeling of ship operations in pack ice. In *International Conference and Exhibition on Performance of Ships and Structures in Ice*, pages 20–23, Banff Alberta, Canada, 2012. Society of Naval Architects and Marine Engineers.

39. P.E. Danielsson. Euclidean distance mapping. *Computer Graphics and Image Processing*, 14(3):227–248, 1980.

40. H. Deng and D.A. Clausi. Unsupervised segmentation of synthetic aperture radar sea ice imagery using a novel markov random field model. *IEEE Transactions on Geoscience and Remote Sensing*, 43(3):528–538, 2005.

41. F. Devernay and O. Faugeras. Automatic calibration and removal of distortion from scenes of structured environments. In *SPIE's 1995 International Symposium on Optical Science, Engineering, and Instrumentation*, pages 62–72. International Society for Optics and Photonics, 1995.

42. K.B. Eom. Fuzzy clustering approach in unsupervised sea-ice classification. *Neurocomputing*, 25(1):149–166, 1999.

43. F.M. Fetterer, D. Gineris, and R. Kwok. Sea ice type maps from Alaska synthetic aperture radar facility imagery: an assessment. *Journal of Geophysical Research: Oceans*, 99(C11):22443–22458, 1994.

44. H. Finch. Comparison of distance measures in cluster analysis with dichotomous data. *Journal of Data Science*, 3(1):85–100, 2005.

45. H. Freeman and L.S. Davis. A corner-finding algorithm for chain-coded curves. *IEEE Transactions on Computers*, 26(3):297–303, 1977.

46. A.M. Ghalib and R.D. Hryciw. Soil particle size distribution by mosaic imaging and watershed analysis. *Journal of Computing in Civil Engineering*, 13(2):80–87, 1999.

47. C. Gignac, Y. Gauthier, J.S. Bédard, M. Bernier, and D.A. Clausi. High resolution RADARSAT-2 SAR data for sea-ice classification in the neighbourhood of Nunavik's marine infrastructures. In *Proceedings of the 21st International Conference on Port and Ocean Engineering under Arctic Conditions*, Montréal, Canada, 2011.

48. R.C. Gonzalez and R.E. Woods. *Digital Image Processing*. Pearson Prentice Hall, Upper Saddle River, NJ, USA, 2008.

49. R.C. Gonzalez, R.E. Woods, and S.L. Eddins. *Digital Image Processing Using MATLAB*. Pearson Prentice Hall, Upper Saddle River, NJ, USA, 2004.

50. V. Grau, A.U.J. Mewes, M. Alcaniz, R. Kikinis, and S.K. Warfield. Improved watershed transform for medical image segmentation using prior information. *IEEE Transactions on Medical Imaging*, 23(4):447–458, 2004.

51. A. Gürtner, B. Bjørnsen, T.H. Amdahl, S.R. Sberg, and S.H. Teigen. Numerical simulations of managed ice loads on a moored Arctic drillship. In *Proceedings of OTC Arctic Technology Conference*, Houston, Texas, USA, 2012. Offshore Technology Conference.

52. A. Haase, S. van der Werff, and P. Jochmann. DYPIC - dynamic positioning in ice: first

phase of model testing. In *ASME 31st International Conference on Ocean, Offshore and Arctic Engineering*, pages 487–494. American Society of Mechanical Engineers, 2012.

53. O. Hall, G.J. Hay, A. Bouchard, and D.J. Marceau. Detecting dominant landscape objects through multiple scales: an integration of object-specific methods and watershed segmentation. *Landscape Ecology*, 19(1):59–76, 2004.

54. R.J. Hall, N. Hughes, and P. Wadhams. A systematic method of obtaining ice concentration measurements from ship-based observations. *Cold Regions Science and Technology*, 34(2):97–102, 2002.

55. Y. Hara, R.G. Atkins, R.T. Shin, J.A. Kong, S.H. Yueh, and R. Kwok. Application of neural networks for sea ice classification in polarimetric SAR images. *IEEE Transactions on Geoscience and Remote Sensing*, 33(3):740–748, 1995.

56. R.M. Haralick and K. Shanmugam. Textural features for image classification. *IEEE Transactions on Systems, Man, and Cybernetics*, 3(6):610–621, 1973.

57. R. Hartley and A. Zisserman. *Multiple View Geometry in Computer Vision*. Cambridge University Press, 2003.

58. J. Haugen, L. Imsland, S. Løset, and R. Skjetne. Ice observer system for ice management operations. In *Proceeding of the 21st International Ocean and Polar Engineering Conference*, Maui, Hawaii, USA, 2011.

59. D. Haverkamp and C. Tsatsoulis. Information fusion for estimation of summer MIZ ice concentration from SAR imagery. *IEEE Transactions on Geoscience and Remote Sensing*, 37(3):1278–1291, 1999.

60. G.J. Hay, T. Blaschke, D.J. Marceau, and A. Bouchard. A comparison of three image-object methods for the multiscale analysis of landscape structure. *ISPRS Journal of Photogrammetry and Remote Sensing*, 57(5):327–345, 2003.

61. L. He, Z. Peng, B. Everding, X. Wang, C.Y. Han, K.L. Weiss, and W.G. Wee. A comparative study of deformable contour methods on medical image segmentation. *Image and Vision Computing*, 26(2):141–163, 2008.

62. V.J. Hodge and J. Austin. A survey of outlier detection methodologies. *Artificial Intelligence Review*, 22(2):85–126, 2004.

63. Q.A. Holmes, D.R. Nuesch, and R.A. Shuchman. Textural analysis and real-time classification of sea-ice types using digital SAR data. *IEEE Transactions on Geoscience and Remote Sensing*, (2):113–120, 1984.

64. N. Hughes. Sea ice type classification from multichannel passive microwave datasets. In *IEEE International Geoscience and Remote Sensing Symposium*, volume 3, pages III–125. IEEE, 2009.

65. A.K. Jain. Data clustering: 50 years beyond K-means. *Pattern Recognition Letters*, 31(8):651–666, 2010.

66. A.K. Jain and R.C. Dubes. *Algorithms for Clustering Data*. Prentice-Hall, Inc., 1988.

67. R. Jain, R. Kasturi, and B.G. Schunck. *Machine Vision*, volume 5. McGraw-Hill, New York, NY, USA, 1995.

68. N.A. Jenssen, T. Hals, P. Jochmann, A. Haase, X. dal Santo, S. Kerkeni, O. Doucy, A. Gürtner, S.S. Hetschel, P.O. Moslet, I. Metrikin, and S. Løset. DYPIC - a multinational R&D project on DP technology in ice. In *Proceedings of the Dynamic Positioning Conference*, Houston, USA, 2012. Marine Technology Society.

69. S. Ji, H. Li, A. Wang, and Q. Yue. Digital image techniques of sea ice field observation in the Bohai sea. In *Proceedings of the 21st International Conference on Port and Ocean Engineering under Arctic Conditions*, Montréal, Canada, 2011.

70. O.M. Johannessen, V. Alexandrov, I.Y. Frolov, S. Sandven, L.H. Pettersson, L.P. Bobylev, K. Kloster, V.G. Smirnov, Y.U. Mironov, and N.G. Babich. *Remote Sensing of Sea Ice in the Northern Sea Route: Studies and Applications.* Springer, 2006.

71. A. John and J. Richards. *Remote Sensing Digital Image Analysis: An Introduction.* Springer, 5th edition, 2013.

72. K. Karantzalos and D. Argialas. Improving edge detection and watershed segmentation with anisotropic diffusion and morphological levellings. *International Journal of Remote Sensing*, 27(24):5427–5434, 2006.

73. J. Karvonen. Baltic sea ice concentration estimation based on C-band dual-polarized SAR data. *IEEE Transactions on Geoscience and Remote Sensing*, 52(9):5558–5566, 2014.

74. J.A. Karvonen. Baltic sea ice SAR segmentation and classification using modified pulse-coupled neural networks. *IEEE Transactions on Geoscience and Remote Sensing*, 42(7):1566–1574, 2004.

75. M. Kass, A. Witkin, and D. Terzopoulos. Snakes: active contour models. *International Journal of Computer Vision*, 1(4):321–331, 1988.

76. A. Keinonen and I. Robbins. Icebreaker characteristics synthesis, icebreaker performance models, seakeeping, icebreaker escort. *Icebreaker Escort Model User's Guide: Report prepared for Transport Development Centre Canada (TP12812E)*, 3:49, 1998.

77. S. Kern, L. Kaleschke, and D.A. Clausi. A comparison of two 85-GHz SSM/I ice concentration algorithms with AVHRR and ERS-2 SAR imagery. *IEEE Transactions on Geoscience and Remote Sensing*, 41(10):2294–2306, 2003.

78. R. Keys. Cubic convolution interpolation for digital image processing. *IEEE Transactions on Acoustics, Speech, and Signal Processing*, 29(6):1153–1160, 1981.

79. C. Kongoli, S.A. Boukabara, B. Yan, F. Weng, and R. Ferraro. A new sea-ice concentration algorithm based on microwave surface emissivities application to AMSU measurements. *IEEE Transactions on Geoscience and Remote Sensing*, 49(1):175–189, 2011.

80. R. Korsnes. Quantitative analysis of sea ice remote sensing imagery. *International Journal of Remote Sensing*, 14(2):295–311, 1993.

81. R. Kwok, E. Rignot, B. Holt, and R. Onstott. Identification of sea ice types in spaceborne synthetic aperture radar data. *Journal of Geophysical Research*, 97(C2), 1992.

82. S.U. Lee, S.Y. Chung, and R.H. Park. A comparative performance study of several global thresholding techniques for segmentation. *Computer Vision, Graphics, and Image Processing*, 52(2):171–190, 1990.

83. B. Leroy, I.L. Herlin, and L.D. Cohen. Multi-resolution algorithms for active contour models. In *12th International Conference on Analysis and Optimization of Systems*, pages 58–65. Springer, Paris, France, 1996.

84. C. Li, J. Liu, and M.D. Fox. Segmentation of external force field for automatic initialization and splitting of snakes. *Pattern Recognition*, 38(11):1947–1960, 2005.

85. K. Li, Z. Lu, W. Liu, and J. Yin. Cytoplasm and nucleus segmentation in cervical smear images using radiating GVF snake. *Pattern Recognition*, 45(4):1255–1264, 2012.

86. A.K. Liu, S. Martin, and R. Kwok. Tracking of ice edges and ice floes by wavelet analysis of SAR images. *Journal of Atmospheric and Oceanic Technology*, 14(5):1187–1198, 1997.

87. D. Liu and J. Yu. Otsu method and K-means. In *9th International Conference on Hybrid Intelligent Systems*, volume 1, pages 344–349, Shenyang, China, 2009. IEEE.

88. F. Liu, B. Zhao, P.K. Kijewski, L. Wang, and L.H. Schwartz. Liver segmentation for

CT images using GVF snake. *Medical Physics*, 32(12):3699–3706, 2005.

89. H. Liu, H. Guo, and L. Zhang. SVM-based sea ice classification using textural features and concentration from RADARSAT-2 Dual-Pol ScanSAR data. *IEEE Journal of Selected Topics in Applied Earth Observations and Remote Sensing*, 8(4):1601–1613, 2015.

90. F. Long, H. Peng, and E. Myers. Automatic segmentation of nuclei in 3D microscopy images of C. elegans. In *4th IEEE International Symposium on Biomedical Imaging: From Nano to Macro*, pages 536–539. IEEE, 2007.

91. S. Løset, K.N. Shkhinek, O.T. Gudmestad, and K.V. Høyland. *Actions from Ice on Arctic Offshore and Coastal Structures*. LAN Publishing House, 2006.

92. P. Lu and Z. Li. A method of obtaining ice concentration and floe size from shipboard oblique sea ice images. *IEEE Transactions on Geoscience and Remote Sensing*, 48(7):2771–2780, 2010.

93. P. Lu, Z. Li, Z. Zhang, and X. Dong. Aerial observations of floe size distribution in the marginal ice zone of summer Prydz Bay. *Journal of Geophysical Research: Oceans*, 113(C2), 2008.

94. W. Lu, Q. Zhang, R. Lubbad, S. Løset, and R. Skjetne. A shipborne measurement system to acquire sea ice thickness and concentration at engineering scale. In *Proceedings of OTC Arctic Technology Conference*, St. John's Newfoundland and Labrador, Canada, 2016. Offshore Technology Conference.

95. Z. Lu and T. Tong. The application of chain code sum in the edge form analysis. *China Journal of Image and Graphics*, 7(12):1323–1328, 2002. in Chinese.

96. R. Lubbad and S. Løset. A numerical model for real-time simulation of ship–ice interaction. *Cold Regions Science and Technology*, 65(2):111–127, 2011.

97. R. Lubbad, S. Løset, U. Hedman, C. Holub, and D. Matskevitch. Oden Arctic technology research cruise 2015. In *Proceedings of OTC Arctic Technology Conference*, St. John's Newfoundland and Labrador, Canada, 2016. Offshore Technology Conference.

98. R. Lubbad, S. Løset, and R. Skjetne. Numerical simulations verifying Arctic offshore field activities. In *Proceedings of the 23rd International Conference on Port and Ocean Engineering under Arctic Conditions*, Trondheim, Norway, 2015.

99. D.J.C. MacKay. *Information Theory, Inference and Learning Algorithms*. Cambridge University Press, 2003.

100. J. MacQueen. Some methods for classification and analysis of multivariate observations. In *Proceedings of the 15th Berkeley Symposium on Mathematical Statistics and Probability*, volume 1, pages 281–297, Oakland, CA, USA, 1967.

101. P. Maillard, D.A. Clausi, and H. Deng. Operational map-guided classification of SAR sea ice imagery. *IEEE Transactions on Geoscience and Remote Sensing*, 43(12):2940–2951, 2005.

102. N. Malpica, C. Ortiz de Solorzano, J.J. Vaquero, A. Santos, I. Vallcorba, J.M. Garcia-Sagredo, and F.D. Pozo. Applying watershed algorithms to the segmentation of clustered nuclei. *Cytometry*, 28:289–297, 1997.

103. T. Markus and S.T. Dokken. Evaluation of late summer passive microwave Arctic sea ice retrievals. *IEEE Transactions on Geoscience and Remote Sensing*, 40(2):348–356, 2002.

104. S. Martin, R. Drucker, R. Kwok, and B. Holt. Estimation of the thin ice thickness and heat flux for the Chukchi Sea Alaskan coast polynya from Special Sensor Microwave/Imager data, 1990–2001. *Journal of Geophysical Research: Oceans*, 109(C10), 2004.

105. T. McInerney and D. Terzopoulos. Deformable models in medical image analysis: a survey. *Medical Image Analysis*, 1(2):91–108, 1996.

106. I. Metrikin and S. Løset. Nonsmooth 3D discrete element simulation of a drillship in discontinuous ice. In *Proceedings of the 22nd International Conference on Port and Ocean Engineering under Arctic Conditions*, Espoo, Finland, 2013.

107. I. Metrikin, S. Løset, N.A. Jenssen, and S. Kerkeni. Numerical simulation of dynamic positioning in ice. *Marine Technology Society Journal*, 47(2):14–30, 2013.

108. J. Millan and J. Wang. Ice force modeling for DP control systems. In *Proceedings of the Dynamic Positioning Conference*, Houston, Texas, USA, 2011. Marine Technology Society.

109. K.I. Muramoto, T. Endoh, M. Kubo, and K. Matsuura. Sea ice concentration and floe size distribution in the Antarctic using video image processing. In *IEEE International Geoscience and Remote Sensing Symposium*, volume 1, pages 414–416, Singapore, 1997. IEEE.

110. K.I. Muramoto, K. Matsuura, and T. Endoh. Measuring sea-ice concentration and floe-size distribution by image processing. *Annals of Glaciology*, 18(1):33–38, 1993.

111. M. Nixon and A.S. Aguado. *Feature Extraction & Image Processing*. Academic Press, 2nd edition, 2008.

112. S. Ochilov and D.A. Clausi. Operational SAR sea-ice image classification. *IEEE Transactions on Geoscience and Remote Sensing*, 50(11):4397–4408, 2012.

113. World Meteorological Organization. *WMO Sea-ice Nomenclature: Terminology, Codes, Illustrated Glossary and Symbols*. Number 259. Secretariat of the World Meteorological Organization, 1970.

114. N. Otsu. A threshold selection method from gray-level histograms. *Automatica*, 11(285-296):23–27, 1975.

115. M.J. Paget, A.P. Worby, and K.J. Michael. Determining the floe-size distribution of East Antarctic sea ice from digital aerial photographs. *Annals of Glaciology*, 33(1):94–100, 2001.

116. A. Palmer and K. Croasdale. *Arctic Offshore Enginnering*. World Scientific Publishing Company, 2012.

117. N. Paragios, O. Mellina-Gottardo, and V. Ramesh. Gradient vector flow fast geodesic active contours. In *8th IEEE International Conference on Computer Vision*, volume 1, pages 67–73. IEEE, 2001.

118. S.K. Park and R.A. Schowengerdt. Image reconstruction by parametric cubic convolution. *Computer Vision, Graphics, and Image Processing*, 23(3):258–272, 1983.

119. C.A. Pedersen, R. Hall, S. Gerland, A.H. Sivertsen, T. Svenøe, and C. Haas. Combined airborne profiling over Fram Strait sea ice: Fractional sea-ice types, albedo and thickness measurements. *Cold Regions Science and Technology*, 55(1):23–32, 2009.

120. C.S. Poon and M. Braun. Image segmentation by a deformable contour model incorporating region analysis. *Physics in Medicine and Biology*, 42(9):1833, 1997.

121. W.K. Pratt. *Digital Image Processing: PIKS Inside*. John Wiley & Sons, Inc., New York, NY, USA, 3rd edition, 2001.

122. W.H. Press, S.A. Teukolsky, W.T. Vetterling, and B.P. Flannery. *Numerical Recipes in C*, volume 2. Cambridge University Press, 1996.

123. Q.P. Remund, D.G. Long, and M.R. Drinkwater. Polar sea-ice classification using enhanced resolution NSCAT data. In *IEEE International Geoscience and Remote Sensing Symposium*, volume 4, pages 1976–1978. IEEE, 1998.

124. D. Ren, W. Zuo, X. Zhao, Z. Lin, and D. Zhang. Fast gradient vector flow computation

based on augmented Lagrangian method. *Pattern Recognition Letters*, 34(2):219–225, 2013.

125. M.E. Rettmann, X. Han, C. Xu, and J.L. Prince. Automated sulcal segmentation using watersheds on the cortical surface. *NeuroImage*, 15(2):329–344, 2002.

126. J.F. Rivest, P. Soille, and S. Beucher. Morphological gradients. *Journal of Electronic Imaging*, 2(4):326–336, 1993.

127. R. Ronfard. Region-based strategies for active contour models. *International Journal of Computer Vision*, 13(2):229–251, 1994.

128. A. Rosenfeld and J.L. Pfaltz. Distance functions on digital pictures. *Pattern Recognition*, 1(1):33–61, 1968.

129. D.A. Rothrock and A.S. Thorndike. Measuring the sea ice floe size distribution. *Journal of Geophysical Research: Oceans*, 89(C4):6477–6486, 1984.

130. H. Sagan. *Introduction to the Calculus of Variations*. McGraw-Hill, New York, 1969.

131. M. Sayed, I Kubat, and B. Wright. Numerical simulations of ice forces on the Kulluk: the role of ice confinement, ice pressure and ice management. In *Proceedings of OTC Arctic Technology Conference*, Houston, Texas, USA, 2012. Offshore Technology Conference.

132. M. Sayed, I. Kubat, B. Wright, A. Iyerusalimskiy, A. Phadke, and B. Hall. Numerical simulations of ice interaction with a moored structure. In *International Conference and Exhibition on Performance of Ships and Structures in Ice*, Banff, Alberta, Canada, 2012. Society of Naval Architects and Marine Engineers.

133. B. Scheuchl, I. Hajnsek, and I. Cumming. Sea ice classification using multi-frequency polarimetric SAR data. In *IEEE International Geoscience and Remote Sensing Symposium*, volume 3, pages 1914–1916. IEEE, 2002.

134. F. Ségonne, A.M. Dale, E. Busa, M. Glessner, D. Salat, H.K. Hahn, and B. Fischl. A hybrid approach to the skull stripping problem in MRI. *Neuroimage*, 22(3):1060–1075, 2004.

135. G. Sheng, W. Yang, X. Deng, C. He, Y. Cao, and H. Sun. Coastline detection in synthetic aperture radar (SAR) images by integrating watershed transformation and controllable gradient vector flow (GVF) snake model. *IEEE Journal of Oceanic Engineering*, 37(3):375–383, 2012.

136. M. Shokr, A. Lambe, and T. Agnew. A new algorithm (ECICE) to estimate ice concentration from remote sensing observations: an application to 85-GHz passive microwave data. *IEEE Transactions on Geoscience and Remote Sensing*, 46(12):4104–4121, 2008.

137. M. Shokr and N. Sinha. *Sea Ice: Physics and Remote Sensing*. John Wiley & Sons, 2015.

138. M.E. Shokr. Evaluation of second-order texture parameters for sea ice classification from radar images. *Journal of Geophysical Research: Oceans*, 96(C6):10625–10640, 1991.

139. R. Skjetne, L. Imsland, and S. Løset. The Arctic DP research project: effective station-keeping in ice. *Modeling, Identification and Control*, 35(4):191, 2014.

140. G.D. Smith. *Numerical Solution of Partial Differential Equations: Finite Difference Methods*. Oxford University Press, 1985.

141. L.K. Soh, D. Haverkamp, and C. Tsatsoulis. Separating ice-water composites and computing floe size distributions. In *IEEE International Geoscience and Remote Sensing Symposium*, volume 3, pages 1532–1534. IEEE, 1996.

142. L.K. Soh and C. Tsatsoulis. Texture analysis of SAR sea ice imagery using gray level co-occurrence matrices. *IEEE Transactions on Geoscience and Remote Sensing*,

37(2):780–795, 1999.

143. L.K. Soh and C. Tsatsoulis. Unsupervised segmentation of ERS and RADARSAT sea ice images using multiresolution peak detection and aggregated population equalization. *International Journal of Remote Sensing*, 20(15-16):3087–3109, 1999.

144. L.K. Soh, C. Tsatsoulis, D. Gineris, and C. Bertoia. ARKTOS: An intelligent system for SAR sea ice image classification. *IEEE Transactions on Geoscience and Remote Sensing*, 42(1):229–248, 2004.

145. L.K. Soh, C. Tsatsoulis, and B. Holt. Identifying ice floes and computing ice floe distributions in SAR images. In *Analysis of SAR Data of the Polar Oceans*, pages 9–34. Springer, 1998.

146. P. Soille. *Morphological Image Analysis: Principles and Applications*. Springer, Berlin, Germany, 2004.

147. M. Steele. Sea ice melting and floe geometry in a simple ice-ocean model. *Journal of Geophysical Research: Oceans*, 97(C11):17729–17738, 1992.

148. M. Steele, J.H. Morison, and N. Untersteiner. The partition of air-ice-ocean momentum exchange as a function of ice concentration, floe size, and draft. *Journal of Geophysical Research: Oceans*, 94(C9):12739–12750, 1989.

149. A. Steer, A. Worby, and P. Heil. Observed changes in sea-ice floe size distribution during early summer in the western Weddell Sea. *Deep Sea Research Part II: Topical Studies in Oceanography*, 55(8):933–942, 2008.

150. H.Q. Sun and Y.J. Luo. Adaptive watershed segmentation of binary particle image. *Journal of Microscopy*, 233(2):326–330, 2009.

151. A. Talukder, D.P. Casasent, H.W. Lee, P.M. Keagy, and T.F. Schatzki. Modified binary watershed transform for segmentation of agricultural products. In *Proceedings of SPIE: Precision Agriculture and Biological Quality*, volume 3543, pages 53–64, Boston, MA, USA, 1998. SPIE Press.

152. P.N. Tan, M. Steinbach, and V. Kumar. *Introduction to Data Mining*. Addison-Wesley Longman Publishing Co., Inc., Boston, MA, USA, 1st edition, 2005.

153. J. Tang. A multi-direction GVF snake for the segmentation of skin cancer images. *Pattern Recognition*, 42(6):1172–1179, 2009.

154. J. Tang and S.T. Acton. Vessel boundary tracking for intravital microscopy via multiscale gradient vector flow snakes. *IEEE Transactions on Biomedical Engineering*, 51(2):316–324, 2004.

155. T. Toyota and H. Enomoto. Analysis of sea ice floes in the sea of Okhotsk using ADEOS/AVNIR images. In *Proceedings of the 16th IAHR International Symposium on Ice*, pages 211–217, Dunedin, New Zealand, 2002.

156. T. Toyota, C. Haas, and T. Tamura. Size distribution and shape properties of relatively small sea-ice floes in the Antarctic marginal ice zone in late winter. *Deep Sea Research Part II: Topical Studies in Oceanography*, 58(9):1182–1193, 2011.

157. T. Toyota, S. Takatsuji, and M. Nakayama. Characteristics of sea ice floe size distribution in the seasonal ice zone. *Geophysical Research Letters*, 33(2), 2006.

158. M.A. Treiber. *Optimization for Computer Vision*. Springer, London, 2013.

159. E. Trucco and A. Verri. *Introductory Techniques for 3-D Computer Vision*, volume 201. Prentice Hall, 1998.

160. R. Tsai. A versatile camera calibration technique for high-accuracy 3d machine vision metrology using off-the-shelf TV cameras and lenses. *IEEE Journal on Robotics and Automation*, 3(4):323–344, 1987.

161. M.A. Tschudi, J.A. Curry, and J.A. Maslanik. Determination of areal surface-feature

coverage in the Beaufort sea using aircraft video data. *Annals of Glaciology*, 25:434–438, 1997.

162. G. Vachon, M. Sayed, and I. Kubat. Methodology for determination of ice management efficiency. In *International Conference and Exhibition on Performance of Ships and Structures in Ice*, Banff, Alberta, Canada, 2012. Society of Naval Architects and Marine Engineers.

163. E.H. Van den Berg, A.G.C.A. Meesters, J.A.M. Kenter, and W. Schlager. Automated separation of touching grains in digital images of thin sections. *Computers & Geosciences*, 28(2):179–190, 2002.

164. S. van der Werff, A. Haase, R. Huijsmans, and Q. Zhang. Influence of the ice concentration on the ice loads on the hull of a ship in a managed ice field. In *ASME 31st International Conference on Ocean, Offshore and Arctic Engineering*, pages 563–569, Rio de Janeiro, Brazil, 2012. American Society of Mechanical Engineers.

165. M. Veta, A. Huisman, M.A. Viergever, P.J. van Diest, and J.P. Pluim. Marker-controlled watershed segmentation of nuclei in H&E stained breast cancer biopsy images. In *IEEE International Symposium on Biomedical Imaging: From Nano to Macro*, pages 618–621. IEEE, 2011.

166. L. Vincent. Morphological grayscale reconstruction in image analysis: applications and efficient algorithms. *IEEE Transactions on Image Processing*, 2(2):176–201, 1993.

167. L. Vincent and P. Soille. Watersheds in digital spaces: an efficient algorithm based on immersion simulations. *IEEE Transactions on Pattern Analysis and Machine Intelligence*, 13(6):583–598, 1991.

168. P. Wadhams. *Ice in the Ocean*. Taylor & Francis, 2000.

169. B. Weissling, S. Ackley, P. Wagner, and H. Xie. EISCAM-digital image acquisition and processing for sea ice parameters from ships. *Cold Regions Science and Technology*, 57(1):49–60, 2009.

170. J. Weng, P. Cohen, and M. Herniou. Camera calibration with distortion models and accuracy evaluation. *IEEE Transactions on Pattern Analysis and Machine Intelligence*, 14(10):965–980, 1992.

171. B. Wright. Evaluation of full scale data for moored vessel stationkeeping in pack ice. Technical report, PERD/CHC, 1999.

172. C. Xu. *Deformable Models with Application to Human Cerebral Cortex Reconstruction from Magnetic Resonance Images*. PhD thesis, Johns Hopkins University, 1999.

173. C. Xu, D.L. Pham, and J.L. Prince. Image segmentation using deformable models. *Handbook of Medical Imaging*, 2:129–174, 2000.

174. C. Xu and J.L. Prince. Snakes, shapes, and gradient vector flow. *IEEE Transactions on Image Processing*, 7(3):359–369, 1998.

175. C. Xu, A. Yezzi, and J.L. Prince. On the relationship between parametric and geometric active contours. In *Proceedings of the 34th Asilomar Conference on Signals, Systems and Computers*, volume 1, pages 483–489, Pacific Grove, CA, USA, 2000. IEEE.

176. X. Yang and D.A. Clausi. SAR sea ice image segmentation based on edge-preserving watersheds. In *4th Canadian Conference on Computer and Robot Vision*, pages 426–431, Montréal, Canada, 2007. IEEE.

177. X. Yang and D.A. Clausi. Evaluating SAR sea ice image segmentation using edge-preserving region-based MRFs. *IEEE Journal of Selected Topics in Applied Earth Observations and Remote Sensing*, 5(5):1383–1393, 2012.

178. X. Yang, H. Li, and X. Zhou. Nuclei segmentation using marker-controlled watershed, tracking using mean-shift, and Kalman filter in time-lapse microscopy. *IEEE*

Transactions on Circuits and Systems I: Regular Papers, 53(11):2405–2414, 2006.

179. P. Yu, A.K. Qin, and D.A. Clausi. Unsupervised polarimetric SAR image segmentation and classification using region growing with edge penalty. *IEEE Transactions on Geoscience and Remote Sensing*, 50(4):1302–1317, 2012.

180. Q. Yu and D.A. Clausi. SAR sea-ice image analysis based on iterative region growing using semantics. *IEEE Transactions on Geoscience and Remote Sensing*, 45(12):3919–3931, 2007.

181. N.Y. Zakhvatkina, V.Y. Alexandrov, O.M. Johannessen, S. Sandven, and I.Y. Frolov. Classification of sea ice types in ENVISAT synthetic aperture radar images. *IEEE Transactions on Geoscience and Remote Sensing*, 51(5):2587–2600, 2013.

182. F. Zamani and R. Safabakhsh. An unsupervised GVF snake approach for white blood cell segmentation based on nucleus. In *8th International Conference on Signal Processing*, volume 2. IEEE, 2006.

183. Q. Zhang and R. Skjetne. Image techniques for identifying sea-ice parameters. *Modeling, Identification and Control*, 35(4):293–301, 2014.

184. Q. Zhang and R. Skjetne. Image processing for identification of sea-ice floes and the floe size distributions. *IEEE Transactions on Geoscience and Remote Sensing*, 53(5):2913–2924, 2015.

185. Q. Zhang, R. Skjetne, S. Løset, and A. Marchenko. Digital image processing for sea ice observations in support to Arctic DP operations. In *ASME 31st International Conference on Ocean, Offshore and Arctic Engineering*, pages 555–561, Rio de Janeiro, Brazil, 2012. American Society of Mechanical Engineers.

186. Q. Zhang, R. Skjetne, I. Metrikin, and S. Løset. Image processing for ice floe analyses in broken-ice model testing. *Cold Regions Science and Technology*, 111:27–38, 2015.

187. Q. Zhang, R. Skjetne, and B. Su. Automatic image segmentation for boundary detection of apparently connected sea-ice floes. In *Proceedings of the 22nd International Conference on Port and Ocean Engineering under Arctic Conditions*, Espoo, Finland, 2013.

188. Q. Zhang, S. van der Werff, I. Metrikin, S. Løset, and R. Skjetne. Image processing for the analysis of an evolving broken-ice field in model testing. In *ASME 31st International Conference on Ocean, Offshore and Arctic Engineering*, pages 597–605, Rio de Janeiro, Brazil, 2012. American Society of Mechanical Engineers.

Index

Printed and bound by CPI Group (UK) Ltd, Croydon, CR0 4YY

24/10/2024

01778301-0008